Copyright & Disclaimer Page

Dedication

To Astrid and Isabella —
like all kids, you remind us every day to stay curious, laugh loudly, and keep asking "What if...?"

You are the fizz in the bubbles,
the sparkle in the foam,
and the best explosions of joy a parent could ever hope for.

This book — and every crazy science adventure — is inspired by you.

Publisher:
Crazy Scientist Press, The
An imprint of The Crazy Scientist® Pty Ltd
Sydney, NSW
Australia
ISBN: 978-1-7644240-0-4
Printed in USA

⚠️ Limit of Liability / Disclaimer of Warranty

This book is intended for educational and entertainment purposes only. The authors and publisher make no representations or warranties with respect to the accuracy or completeness of the contents and specifically disclaim all warranties. The experiments and activities are provided with the understanding that the publisher and authors shall not be liable for damages, injury, or loss resulting from the use of this material.

Adult supervision is required for all activities. The authors and publisher disclaim liability for any damage, mishap, or injury that may occur while engaging in the activities featured in this book.

Safety Rules for Junior Scientists

🥽 **Wear your goggles** — eyeballs are precious!

🧑‍🤝‍🧑 **Ask an adult** lab assistant before using scissors, microwaves, or rockets.

🧼 **Keep it clean** — science is more fun when you're not sticky from head to toe.

🚫 **Don't taste the experiments** — unless it's clearly marked as safe food fun.

🧯 **Be prepared** — paper towels and a tray are your best lab buddies.

Remember: The best experiments are the ones you can laugh about safely afterwards!

Contents

Contents

Welcome to the LAB

Where portals open, gadgets misbehave and science gets wildly fun.

You've officially stepped inside the Secret Lab — the place where experiments explode, gadgets misbehave, and portals open to just about anywhere.

Inside this lab sits the Porthole Reactor, a swirling gateway powered by gross gases, giggle energy, and good old-fashioned science.
During your adventure you'll jump from one strange location to another, performing silly, surprising, and sometimes slightly chaotic science experiments.

Ready?
Goggles on. Curiosity activated.
Science adventure... GO!

INSIDE THE WHIFFINATOR™

The Whiffinator™ is a portable stink-science machine. It vacuums odours in, whirls their gas molecules through a turbo-mixer, super-charges the chemical "stink particles," and then blasts them back out at high velocity. Powered by a Dad-Fart Reactor Core, it can recreate anything from Dog-Poo Power to the classic stink-bomb sulphur. Slide the Intake Valve to sniff smells in—or flip the Reverse Blower Switch to fire them out and tickle nearby olfactory nerve endings (a.k.a., nostrils).

Safety Note: If the Meltdown Relief Valve starts giggling, step away immediately.

Weight: 2.5 kg

Power Level: Grossly Powerful

Smell Radius: 1–7 metres

Safety Rating: Highly questionable

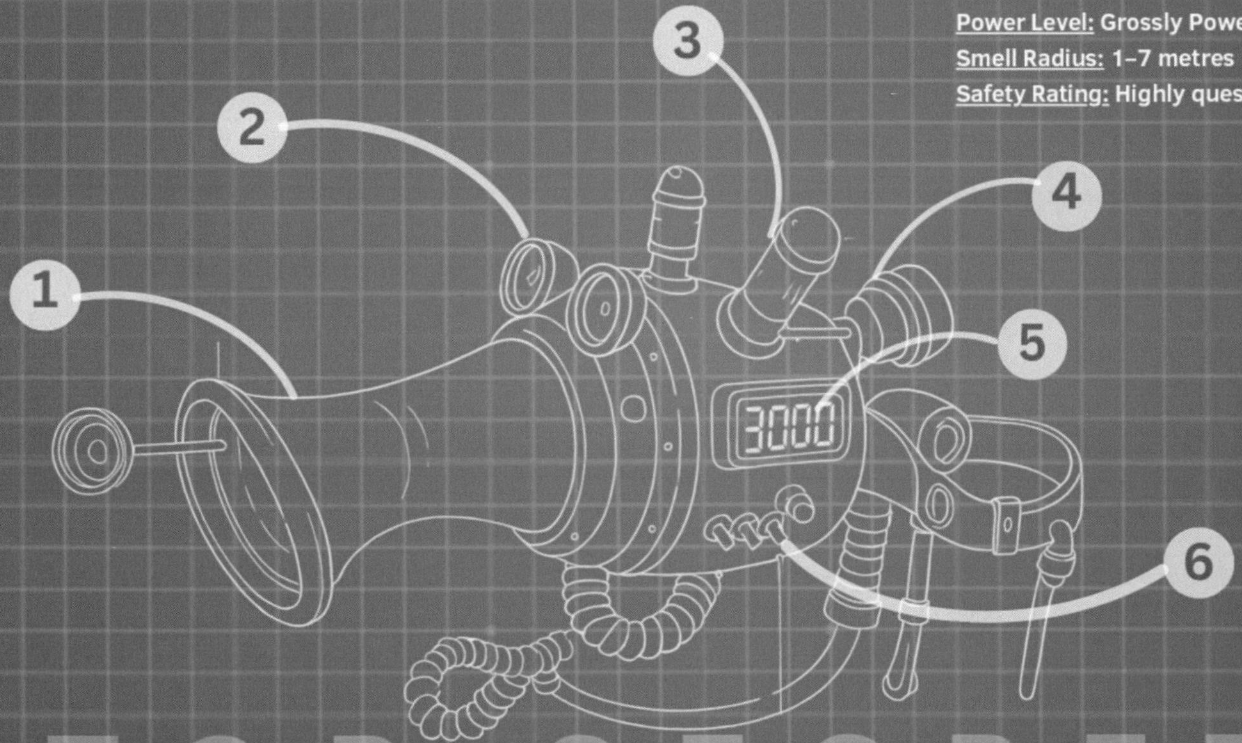

TOP SECRET

1. Turbo-Slurper Intake Valve
2. Glow-O-Meter Tube
3. Nitro-Cool Stabiliser (DO NOT SHAKE)
4. P.U. Perception Lens
5. Gas Particle Peek-O-Scope
6. Chaos Control Cluster

Meet The Gadget Wall!

These are just some of the wild, wacky, wonderfully ridiculous inventions built inside the Secret Lab. Each one has its own job... and its own potential to blow up, fizz, wobble, squeak, or accidentally summon a giant soap bubble.

The Slimometer

The Whiffinator

The Frostinator 6000

The Portal Controller

Submarine

The Shrinkgun

The Soapanator

Spaceship

Welcome to the Secret Lab!

Hey! I'm Dr Puddledrip, one of the crazy scientists who makes amazing experiments in this secret lab.

Professor Picklebottom and I use a science porthole to jump to different places in time (past, present, and future) and try wild, wacky experiments. You can do them too as you read this book!

Sometimes the porthole will light up—when you see the Portal Activated symbol, tighten your goggles... we're jumping!

How we do science (every time):

1. What if... Ask a curious question.
2. Make a guess. Your best idea about what will happen.
3. Test it out. Follow the steps—observe safely.
4. Look closely. What actually happened?
5. Try another way. Change one thing and test again.

A Note for Parents

Curiosity Comes First

As a parent and science educator, I've learned that the things that spark real curiosity in kids are simple: wonder, play, and the freedom to make a glorious mess. Children don't need long explanations to start thinking like scientists. They need the chance to ask "What if...?", the courage to try an idea, and encouragement to look closely at what happens next.

I've seen this in classrooms, shows, and workshops all over Australia. When children are given room to experiment, explore, and get a little messy, that's when the magic happens — and when real learning begins.

More Than Just Recipes

After 30 years of teaching science — and 15 years performing shows for tens of thousands of children — I've noticed something: many experiment books stop at the recipe. Do this, get that. Kids might see the result, but not the idea behind it.

This book works differently.
Every experiment includes simple diagrams and visual explanations that help make invisible ideas visible. Children will see gas forming, pressure building, and energy being released — not just read about it. The goal is to help them understand why something happens, not just what happens.

A Note for Parents

STEAM for Every Child

Every activity in this book includes quick STEAM extensions that let children explore ideas through Science, Technology, Engineering, Art, and Maths. This matters because kids learn in different ways.

- Some respond to numbers and measurement.
- Some connect through drawing and creativity.
- Others understand ideas best by tinkering, building, or testing again.

By offering variety, we give every child a way in — and a chance to say, "I get it!"

Learning together...

You don't have to be a trained scientist to utilise this book. The materials consist of everyday items, and the instructions are straightforward. Your role is to embrace the wonder, pose silly questions, share your guesses, and celebrate the surprises along the way. Learning occurs organically through the process of doing.

And if you'd like to go further, you can join the Secret Lab membership for animated episodes, more experiments, worksheets, and competitions.

You'll also find extra tips on how to modify experiments and dive deeper. There's also a community of curious families exploring science together.

Think of it as your science safety net — you'll always have support when you need it.

Exploring the circulatory system in our mini-sub!

How to Use This Book-TEACHERS:

Behind the fizz and fun...
This book is built on solid science teaching practice. Each experiment is structured around the experimental method, guiding students through:

- Asking a question
- Making a prediction (hypothesis)
- Testing
- Observing and recording results
- Modifying variables and repeating

This approach develops inquiry skills and scientific habits of mind — without the weight of a formal lab report.

Why this works in classrooms

✓ **Multiple learning modalities** — visual diagrams, kinaesthetic hands-on tasks, verbal & written reflection.

✓ **Multiple intelligences** — explore through measurement & numbers (logical-mathematical), creativity & drawing (visual-spatial), building & design (bodily-kinaesthetic), or storytelling & discussion (linguistic-interpersonal).

✓ **Variable manipulation** — built-in opportunities to adjust conditions, compare results and extend inquiry.

✓ **STEAM integration** — activities connect naturally to Technology, Engineering, Art, Maths, as well as Science.

✨ **Tip from a teacher (30 years in classrooms + 15 years running science shows)**

The magic isn't just in the fizz — it's in giving students permission to change variables, explore the "what if," and discover that science is a process, not just an answer.

More Crazy Science Adventures Ahead

This is only the beginning. The Secret Lab Science series will grow into a full universe of hands-on experiments, portal adventures, wild gadgets and STEM stories that explore every corner of science — from bubbling chemistry to crazy forces, lasers, electricity, the human body, frozen worlds, outer space and more.

The goal?
To give kids and classrooms a whole portal-powered world of science they can explore, collect and return to again and again.

Keep your goggles polished and your curiosity switched on — new releases, companion podcasts, Secret Lab Membership resources and adventure updates are on the way!

Join the mailing list for first-release specials, free experiments and early access: www.thecrazyscientist.com

A Message From The Crazy Scientist

Whether you're a parent searching for a Saturday spark, or a teacher looking to bring more joy and hands-on learning into the classroom, this book is for you.

My hope is that these experiments, stories and portal discoveries bring fizzing fun to your home or school — and remind every young scientist that science isn't about knowing everything.

It's about exploring, wondering... and daring to ask bold questions.

— Darin Carr, BScDipEd
The Crazy Scientist Pty Ltd®

A Note for Teachers

If you're reading this as an educator, you'll recognise the scaffolding underneath. Each experiment in this book aligns with the processes of experimental design.

The hidden pattern of question → prediction → test → observe → repeat is deliberately built in. These resources support inquiry-based learning, skill development, and curriculum outcomes in primary science classrooms.

For full classroom support, the Secret Lab Teacher Membership includes:

- Lesson plans that use formal scientific terminology (hypothesis, variables, reliability, and more).
- Worksheets, graph templates, and assessment notes.
- Extensions that link these playful experiments to curriculum outcomes.

Blending Story, Portals + Crazy Science

This book does something unusual for kids' science: it mixes hands-on experiments with a wild, ongoing story universe. Your child won't just follow steps - they'll jump through glowing portals with Professor Picklebottom and Dr Puddledrip, exploring strange worlds while doing real experiments. Every five missions, a new portal glitch or cliffhanger pulls them back into the bigger adventure that runs across the whole series, keeping them excited, imaginative, and feeling part of a connected world of discovery.

Inside this book, your child taps into two superpowers:

The Power of Story

Stories pull kids into learning. They spark imagination, build emotional engagement, and keep pages turning.

The Secret Lab universe ties together:

- the book series
- animations on YouTube & The Secret Lab
- audio podcasts

and even future young adult novels like Mackey Maybel and the Time-Ripple Mysteries

The Power of Hands-On Science

- Experiments give kids the thrill of doing.
- They're messy, surprising, and confidence-boosting.
- Each mission feels like a real challenge kids can complete — not just read about.

Why Combine Them?

Because when kids feel like they're part of the Secret Lab story...they're:

- more curious
- more motivated
- more confident
- and more excited to keep exploring

It's reading + science + imagination —

a complete Universe of Crazy Science discovery.

How to Use This Book – KIDS

1. Make a guess
2. Test your idea.
3. What happened?
4. Try it again in a new way.

Have fun & explore

How to Use This Book – PARENTS:

You don't need to be a scientist to use this book.

The materials use everyday items, the steps are clear, and the fun is built in. Your role is simply to:

Ask the silly questions. Share the guesses. Celebrate the mess and surprises.

Behind the scenes...
Each experiment secretly follows the steps of the scientific method (good science):

- Asking a question (What if...?)
- Making a prediction (a guess!).
- Testing it out
- Looking closely at the result
- Trying it again in a new way

This way, kids learn the habits of real scientists — without feeling like they're in a classroom.

What makes this book different?

- Visual explanations — to make the hidden "why" come alive.
- STEAM extensions — each experiment includes simple ways to expand into Science, Technology, Engineering, Art, or Math
- Modifications — built-in ideas to change variables, test new angles, and dive deeper.

Tip from a teacher (30 years in classrooms + 15 years running shows):

The best learning doesn't happen when things go "right." It happens when kids shout "Wait – why did that happen?!"

The Secret Lab Science Series

Step inside the Secret Lab... where swirling portals, strange planets, and mind-bending gadgets turn every experiment into an adventure. The Secret Lab Science Series blends real-world STEM with fun, fast-paced stories that transport kids across the universe while teaching the science behind each discovery. From volcano blasts to laser mazes, sneezing snot monsters to frozen worlds — every book becomes a new portal to explore.

Pick your favourite topic below and jump into the adventure!

Book 1 — Science Experiments for Kids

Book 2 — Earth Unzipped

Book 3 — Sonic Science, Slime & Gross Chemistry

Book 4 — Forces, Flight & Engineering Experiments

Book 5 — Light & Lasers in the Luminous Labyrinth

Book 6 — Shocking Science in the Storm Tower

Book 7 — Space Science & Cosmic Forces

Book 8 — Bogus Biology: The Human Body, Microbes & More

Book 9 — Frozen & Fantastic Science

Book 10 — Magnet Mayhem, Forces & Fantastic Physics

...AND MORE PORTAL-POWERED ADVENTURES COMING SOON

Choose Your Science Adventure!

Kids' Favourite Experiment Topics	Then you will find them in this book...
Explosions, volcanoes, fizzing reactions	Book 2
Slime, snot, gross science	Book 3, Book 8
Forces and flight	Book 4
Light, lasers, colour, optics	Book 5
Electricity, circuits, shocking reactions	Book 6
Space, rockets, astronomy, the Moon	Book 7
Ice experiments, freezing, states of matter	Book 9
Human body science (pulse, lungs, senses)	Book 8, Book 1
Magnets & magnetic fields	Book 10
Chemistry basics (solubility, colour changes, simple reactions)	Book 1
Bubbles, surface tension, water tricks	Book 1
Sound science (pitch, vibration, weird noises)	Book 3, Book 1
Simple machines, engineering challenges	Book 4, Book 7
Weather experiments (wind, clouds, storms)	Book 6
Biology / life science (plants, germs, living things)	Book 8, Book 1
Portal & story-based science adventures	Every book in series

For Teachers & Schools

Every book in the Secret Lab Science Series connects directly to the Teacher Secret Lab Membership — your all-in-one hub for curriculum-aligned, ready-to-teach resources based on the NSW Science & Technology K–6 Syllabus (2024).

Inside the Teacher Secret Lab you'll find:

- Curriculum-aligned worksheets (NSW syllabus outcomes clearly mapped)
- Complete lesson plans & sequencing guides (editable + printable)
- Experiment sheets & materials lists (classroom-friendly, low prep)
- Assessment guides, rubrics & differentiation ideas
- Digital portal stories + animations to bring each book to life
- Extension challenges & classroom posters
- Collector card sets (30 per book) for stations, rewards & inquiry tasks

Perfect for:

K–6 classrooms, science clubs, enrichment groups, STEM days, relief-teacher lessons, and cross-stage units.

CHECK SYLLABUS COVERAGE

Aligned to the 2024 NSW Science & Technology K-6 Syllabus
Every book in the Secret Lab Science Series has been carefully mapped to the strands and outcomes in the 2024 NSW Science & Technology K-6 syllabus.

Working Scientifically → ALL books
Developing inquiry, observation, prediction, investigation, design & communication skills.

Living World → Books 8 + 1
Human body, microbes, plants, growth, life cycles.

Material World → Books 1, 3, 9
Mixing, solubility, changes of state, reversible & irreversible changes.

Physical World → Books 3, 4, 5, 6, 9, 10
Forces, flight, sound, light, electricity, magnetism, motion, energy.

Earth & Space → Books 2 + 7
Earth's surface, natural events, the Moon, planetary movement, environmental interactions.

Digital Technologies → Integrated across the series
Especially Book 4 and Book 6 (design thinking, circuits, tech systems).

Design & Engineering → Books 4 + 7
Planning, creating, testing, and improving designs & models.

Time for Experiments!

Professor Picklebottom bursts through a side door and yells:
"Dr. Puddledrip! The Portal Reactor is humming again! That means only one thing — IT'S TIME FOR EXPERIMENTS!"

Dr. Puddledrip adjusts her goggles and taps her clipboard.
"Calm down, Professor. We need to tell our junior scientists how it all works first. After all, every experiment in this book works like a mini-mission."

So, little scientists, you'll discover the science by doing the experiments... and each result helps stabilise the portals."

Picklebottom gasps, "Wait, what?"

"You hear that, kids? YOUR experiments power our portal jumps! Mix something, test something, measure something... and WHOOSH — the portal takes us somewhere new!"

Dr. Puddledrip points toward the glowing Gadget Wall.
"Some portals may require a special device — the Shrinkascope™, the Whiffinator 3000™, or the Sub-Sea Explorer Pod... so check the blueprints at the beginning of the book."

Picklebottom leans in and whispers dramatically:
"And remember... not all portals are predictable. Some lead to planets full of slime, some lead to windy mountaintops. One even dropped us inside a giant dinosaur sneeze. But we don't talk about that one."
Puddledrip sighs. "Yeah, we talk about it WAY too often! In fact, I think you love snot, poo and vomit way too much."
Picklebottom claps his hands while smiling and laughing.
"Anyway! Ready to power up the Portal Reactor?"
The room begins to tremble...

Portal Activation: Mission #1 Begins!

The Portal Reactor flickers once... twice... then erupts in a blinding flash of yellow.

Green.

Purple.

KA-THOOM!

Dr Puddledrip checks the swirling readings on her tablet.
"According to this — the first portal jump depends on our first experiment.

Picklebottom's goggles fog up with excitement.
"So what is it this time... bubbles? Rockets? Lightning? Frozen science?
Wherever you start — WE go!"

A trembling ring of blue light opens in the centre of the room. The air crackles.
Loose papers flutter. The Gadget Wall gives a low, suspicious hum, like it's thinking...

GADGET WALL ALERT!

Will we need a device?

A scanner?

A gizmo?

Puddledrip notices a gadget starting to glow.
It shouldn't glow... but it definitely does.

"Keep your eyes sharp, scientists. You never know which gadget gets selected."

Dr Puddledrip steps forward — steady, but smiling.
"Junior scientists, your first experiment is on the next page.
Follow the steps carefully... the portal responds to your results."

Picklebottom waves frantically.
"Grab your lab coats! Tighten your goggles!
The adventure begins the moment you turn the page!"

The floor vibrates...
The lights flicker...
The portal swells, glowing like a giant heartbeat...

WHOOOM... WHOOOM... WHOOOM — Portal Jump incoming... hold on tight... here we goooooo!

Boogie Bubbles & Raisin Rave!

Hold onto your lab coats, junior scientists — these wrinkly little raisins are about to throw the fizziest, floatiest, bubble-blasting boogie bash this side of the fruit bowl!

Portal Progress

Attention, curious scientists! I have here a glass of mysterious fizzy potion and a handful of wrinkly raisins. What will happen when the two meet? Will they plummet to the depths? Leap for joy? Or do something so bizarre that it makes your eyebrows dance? There's only one way to find out — let's experiment!

"Before we start — what do YOU think will happen to the raisins? Will they sink like a stone, float like a boat, or dance like disco champions?"

What You Need

- Two x plastic glasses (champagne flute or tall container)
- Raisins (sultanas & small pasta)
- Fizzy drink

Try It Another Way!

"Swap raisins for popcorn kernels, small pasta shells, or dried cranberries. Which ones dance best?

Step-by-Step:

1. Set the Stage

- Place two clear cups or tall glasses on the table.
- Label one "Fizz Fest" (for soda water) and one "Still Water".

2. Pour the Potions

- Fill the "Fizz Fest" glass with soda water right to the top. Fill "Still water" with the same amount of still water.

3. Make Your Predictions

- Ask: What will the raisins do in each glass? Will they sink? Float? Dance like they just won a talent show?
- Write down or draw your predictions.

4. Drop the Raisins

- Plop a few raisins into each glass one at a time.
- Watch closely — the action can start fast!

5. Observe the Dance Floor

- Time how long it takes for the first raisin to rise in the fizzy drink. Compare what's happening in the still water vs. the fizzy water. Which glass has the best dance moves?

Professor Picklebottom's Bubble Boogie Breakdown

When soda water is made, carbon dioxide gas (CO_2) is pushed into the liquid under high pressure. When you open the bottle — SHHHH! — some of that gas escapes right away, but plenty remains dissolved in the drink. The gas is itching to escape... but it needs somewhere to begin.

That's where the raisin comes in! Its wrinkly, bumpy skin is covered in tiny pits and crevices. These are called nucleation sites — perfect little spots where dissolved CO_2 can turn into bubbles. You can see the same thing on the sides of a glass, where tiny scratches or imperfections make bubbles appear.

Once bubbles form on the raisin, they act like dozens of mini life jackets, making the raisin more buoyant and helping it rise toward the surface.

But wait — there's even more bubble magic! Those same bumps can grab CO_2 bubbles already floating in the soda. As bubbles bump into the raisin, they stick to its rough skin, giving it an extra lift. When the bubbles reach the surface — POP! — the gas escapes, and the raisin sinks back down.

Then the dance starts all over again... until the soda goes flat and the party's over!

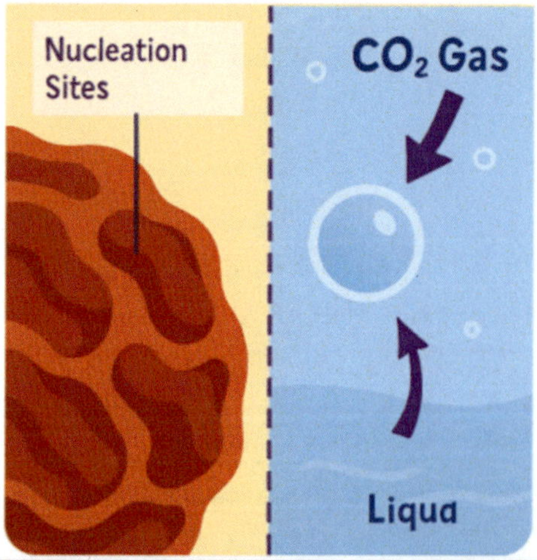

Take it Further with Exclusive Parent & Teacher Resources

Next-Level Science: Join the Secret Lab membership. I'll show you how to make it a chemical change too!

You'll get:

- Safe chemical reaction versions of this trick.
- Hypothesis, variables & more.
- Teacher notes comparing physical vs chemical changes.

www.thecrazyscientistlab.com.au

S Investigate different liquids to see which makes the sultanas dance best. Try soda water, lemonade, sparkling mineral water, and cola. Record which one produces the most dancing and longest activity.

T Research and compare how different drink bottling methods trap gas. Use a camera or phone to take time-lapse footage of the sultanas dancing, then analyse the video to see bubble patterns and timing.

E Design and build a "Dance Arena" — a custom container (clear plastic tube, tall vase, or stacked bottles) that makes the sultanas bounce higher or longer. Experiment with different shapes, heights, and widths.

A Create a comic strip or short stop-motion animation starring your sultanas. Give them names, facial expressions, and dialogue as they dance, collide, and fall back to the bottom.

M Count the number of bounces each sultana makes in one minute. Make a bar graph or line graph comparing different sultanas or different liquids. Calculate averages and find out which liquid gives the highest bounce count.

Mission 02

Soda Supernova

Ever seen a soda bottle shoot a foamy fountain into the sky? Today, you'll discover what happens when chemistry, bubbles, and a little candy team up for a fizzy explosion.

Portal Progress

Spoiler alert: It's messy, it's loud, and it's crazy fun — so let's blast off! Ever seen a fountain shoot sky-high like a rocket?

Why does this happen? Surface tension, carbon dioxide bubbles, and Mentos magic team up for a show-stopping explosion.

Before we start — what do you think will happen when Mentos meets soda? Which soda will make the biggest eruption — diet, regular, or sparkling water? Make your predictions!

What You Need

- Soft drink bottles (Diet Coke, etc.)
- Mentos tablets (other flavours optional if testing)
- Paper, cardboard, tape & scissors
- Safety glasses

Try It Another Way!

How is the gas production and geyser height affected when you use different flavours of Mentos?

Step-by-Step:

1. Make the Tube

- Roll the A4 paper into a cylinder just wide enough for Mentos to fit inside.

2. Secure the Tube

- Tape the top edge, bottom edge, and one long strip along the seam so the tube holds its shape.

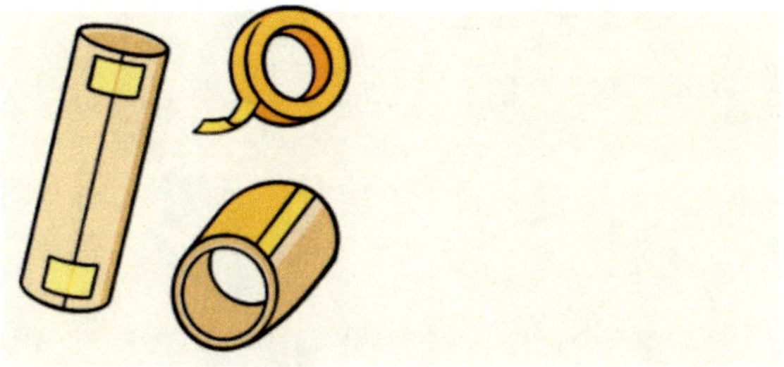

3. Position the Tube

- Slide the empty paper tube so it rests just outside the mouth of the soda bottle.
- Tape it in place firmly so it doesn't move.

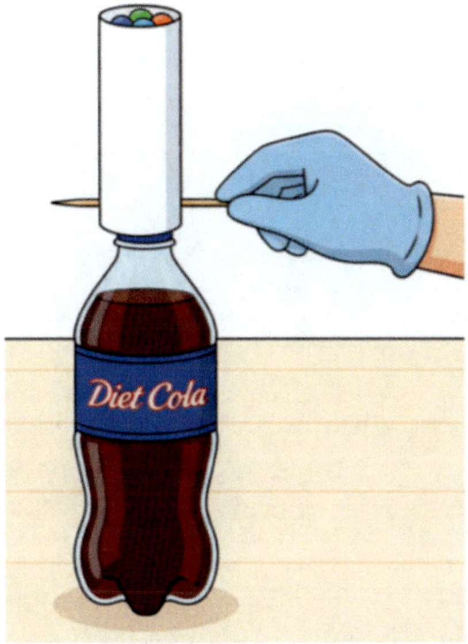

4. Add the Mentos Stopper

- Pierce a toothpick through the bottom of the tube (near the mouth of the bottle).
- This acts as a stopper so the Mentos don't fall in.

5. Load the Mentos

6. Stand Back and Launch

- Hold the bottle upright, pull the toothpick, and watch your geyser blast sky-high!

Professor Picklebottom's Breakdown

When Coke is made, carbon dioxide gas (CO_2) is squeezed into the liquid under very high pressure. It stays hidden, all squished inside the drink, until you open the bottle — SHHHHH! Then the gas wants to escape.

But it needs a place to start. That's where the Mentos come in!

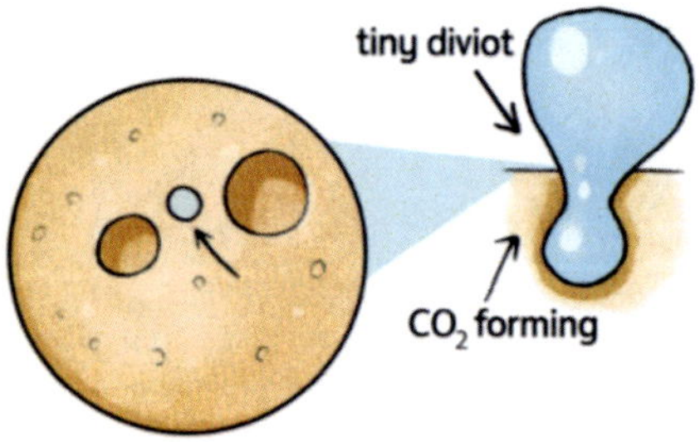

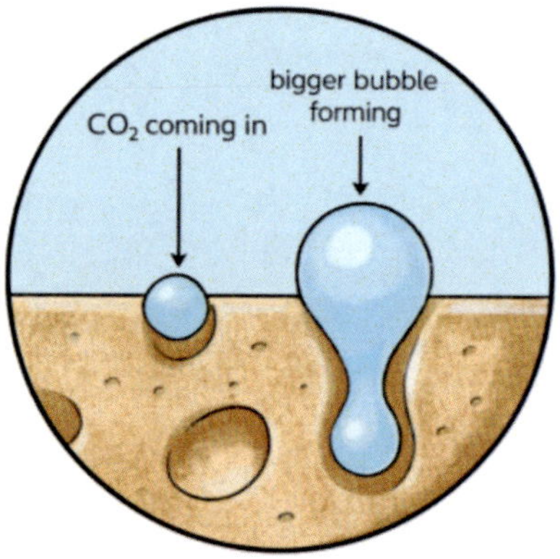

On the surface of a Mentos tablet are hundreds of tiny scratches and holes. These rough spots give the carbon dioxide gas (CO_2) in the Coke lots of places to start forming bubbles. Each one of these spots is called a nucleation site.

A nucleation site is like a bubble starting point. The more sites there are, the more bubbles can form at the same time. As soon as the Mentos drop into the Coke, gas races to these sites, creating hundreds of bubbles all at once.

Each bubble keeps growing bigger because more and more gas joins in. Gas takes up much more space than the liquid it comes from, so all those bubbles push the liquid out of the way. With so many bubbles forming so fast, the pressure inside the bottle builds up quickly—until the only way out is up and out of the bottle in a giant fizzy geyser!

Next-Level Experiments:

- Change the number of Mentos: Does more always mean a higher eruption?
- Test different sodas: Which one rockets the highest?
- Try nozzle control: How does a smaller opening change the spray?

You'll learn about:

✓ Variables and fair testing

✓ Making predictions and recording results

✓ How to turn a fizzy blast into real science!

Experiment with different types of soda (diet, regular, fruit-flavoured) to see which produces the tallest geyser. Record your measurements and compare the results.

Use slow-motion video on a smartphone to capture the geyser eruption. Analyse the video frame by frame to measure its speed, height, and duration.

Design a simple contraption to tilt your RocketFizz Tube at different angles (like 30°, 45°, or 60°) and secure the bottle safely. Test which angle makes the spray travel the farthest or create the coolest arc. Record your results and compare them!

Create a 'RocketFizz Launch Poster' or a fun comic showing the countdown, blast-off, and soda explosion. Add colourful rockets, fizz clouds, and dramatic sound words like BOOM! and WHOOSH!

Graph your results by plotting the number of Mentos versus geyser height. Create bar charts for different soda brands. Calculate averages to see which combination performs best.

Kaboom-in-a-Bag

Have you ever wondered what happens when an acid and a base get together for a little chat? They create a lot of gas, that's what! Today, we're going to trap that gas in a bag and see what happens when the pressure builds. It's a classic chemical caper, and you're the masterminds.

Portal Progress

Before we start – what do YOU think will happen when the paper pouch gets all soggy inside the bag? Will it POP, FIZZ, or just... get a little wet?

What You Need

- A clear, small zip-lock bag (the sturdy kind, please!)
- Baking soda (Professor Picklebottom insists on calling it "bicarb" for extra fanciness)
- White vinegar
- A sheet of tissue paper or paper towel
- A measuring cup
- A spoon
- Safety goggles

Try It Another Way!

Test different bag sizes or different amounts of baking soda and vinegar. Which combination makes the bag inflate quickest? Which one gives the loudest POP?

Step-by-Step:

1. Preparation (The 'Perilous Pouch'):

- Place a teaspoon of baking soda in the centre of your tissue paper. Carefully fold the paper into a small but loose bundle, like a secret science package.

2. The 'Vexing Vinegar':

- Using your measuring cup, pour about 1/3 cup of white vinegar into your zip-lock bag.
- Try not to splash it about!

3. The 'Grand Gauntlet':

- This is the tricky part, so be quick! Hold the bag open and, using your non-dominant hand, drop the baking soda pouch into the bag.

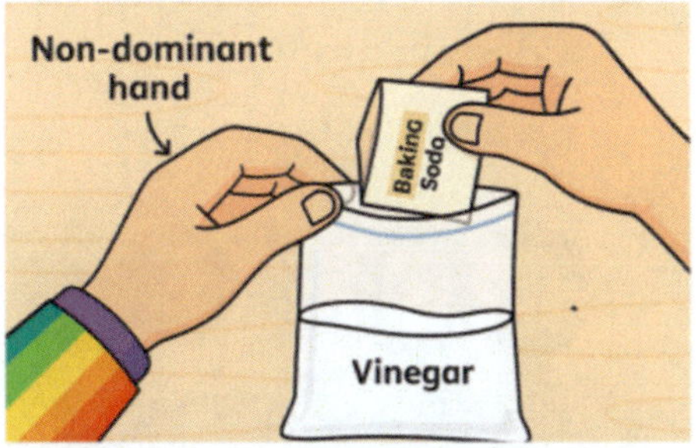

4. The 'Secure Seal':

- Quickly zip the bag shut.
- Make sure there are no leaks!

5. The 'Great Retreat':

- Shake it a little to help the powder come out, then place the bag in your tray or on the ground outside... and stand back!

Watch the bag inflate and listen closely for the 'POP!'

Professor Picklebottom's "Boom Boom Chemistry" Breakdown

When you mix bicarbonate of soda (a base) with vinegar (an acid), a chemical reaction takes place.
It produces a brand-new gas: carbon dioxide (CO_2).
At first, the gas has nowhere to go, so it starts filling the bag, bumping and jostling around inside.
As more and more gas is produced, the bag swells like a puffer fish.
The pressure keeps building... until the bag can't hold it any longer.
That's when you hear the big POP!

Carbon dioxide molecules

A carbon dioxide molecule has:

1 carbon atom
2 oxygen atoms

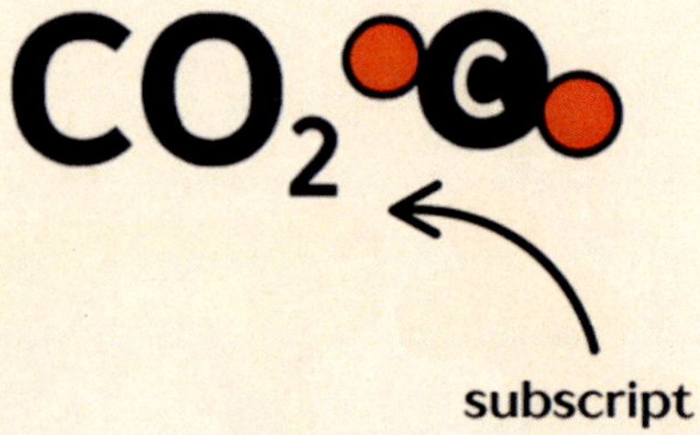

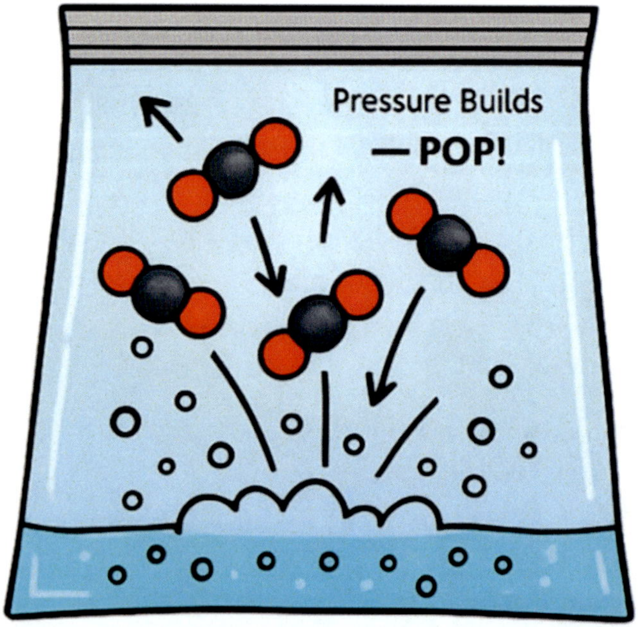

Did you know?

The gas pushes on every side of the bag. When there's too much pressure, the bag can't stretch any further – so it *explodes*!

What is a subscript?

A subscript is a small number written low and to the right of a chemical symbol. It tells you how many atoms of that element are in the molecule.

Take It Further with Your Kaboom Bag!

"Want to turn your Kaboom Bag into a real science mission? You can test, tweak, and compare results to discover what makes the biggest POP!

Next-Level Experiments:

- Change the amount of baking soda: Does more powder mean a louder or a faster pop?
- Change the temperature of the vinegar: Does warm vinegar or cold vinegar react faster?
- And much more...

You'll learn about:

✓ Variables and fair testing

✓ Making predictions and recording results

✓ How to turn a fizzy blast into real science!

Test different amounts of baking soda and record how long it takes for the bag to pop.

Could you use a pressure sensor or a phone app to measure how much pressure builds up inside the bag before it "pops"? How would you design a gadget to do that?

Can you build a "stand" or a "launch pad" for your bag that will catch all the vinegar and make the experiment less messy? Use craft sticks, cardboard, or anything you can find!

Use paints and a canvas to create an abstract painting that represents the colours (or colors), movement, and energy of your "Kaboom" explosion!

Does using more baking soda make the bag pop faster? Design a proper experiment to test this. What variables would you need to measure (e.g. time until the bag pops, amount of powder)? Graph your results!

Mission 04

Fizz-tastic Flying Fiasco!

Dr Puddledrip here, and today we're going to make something go up, up, up… and AWAY! Who needs a real rocket when you've got a fizzy tablet and a tiny pot? Let's make some bubbly chaos together!

Portal Progress

We're going to put that fizz to the test! By trapping bubbles inside a small container, we'll build up pressure until—WHOOSH!—our tiny rocket shoots into the sky. It's a "fizzics" challenge, if you will!

Before we start — what do YOU think will make our rocket fly higher? More water? More tablets? Make your prediction… then watch the bubbly action!

What You Need

- An empty film canister with a tight-fitting lid (or any small container with a well-sealing lid)
- Water
- Aspro-Clear or other effervescent tablets (the fizzy ones!)
- Safety goggles
- A safe outdoor space (a grassy area works best!)

Try It Another Way!

What happens if you use different liquids? Try orange juice or lemonade instead of water. Which one launches the rocket the highest?

Step-by-Step:

1. The Preparation:

- Take one effervescent tablet and break it into four pieces.
- This helps it dissolve faster for a quicker reaction!

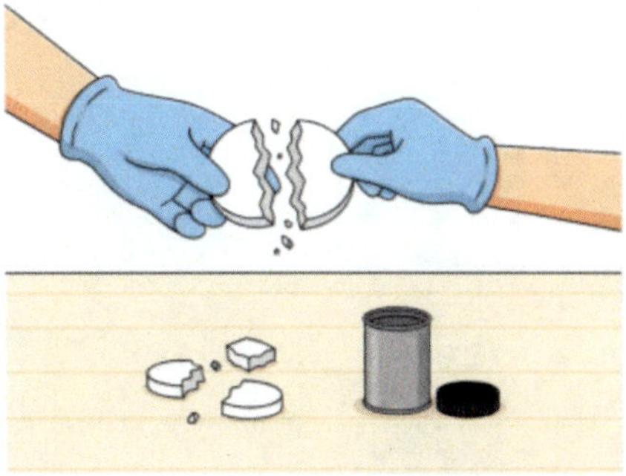

2. The 'Wet and Wild' Chamber:

- Carefully pour about one tablespoon of water into the film canister. Don't fill it up all the way —it needs room for the bubbles to build up!

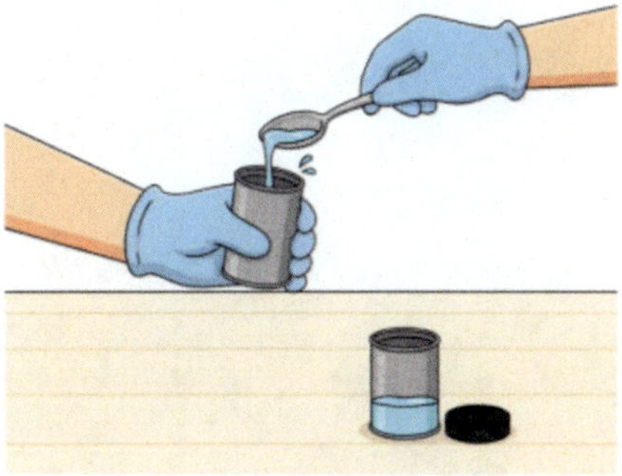

3. The 'Drop-in':

- Drop one of those smaller pieces into the water-filled canister.

4. The 'Seal and Retreat!':

- Immediately and securely snap the lid onto the canister.
- Turn the canister upside down (lid-side (down) and place it on the ground with a straight arm, then move your head and body away from the canister.

5. The 'Ready, Set...POP!':

- Step back and watch your little rocket take off!
- The pressure from the fizzing gas will build up until the lid pops off, sending the canister flying!

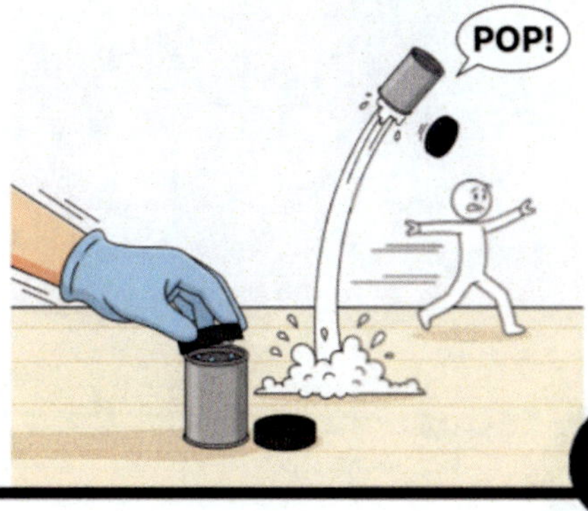

Dr Puddledrip's Chemistry Breakdown

When the tablet is in the water, the two chemicals in the tablet dissolve. This allows the acid and the base to combine and produce a gas called carbon dioxide (CO_2). As the amount of gas increases inside the container, it pushes against all the sides of the container, increasing the pressure.

Since the lid is the weakest point, it will be the first part to give way to the gas pressure. The container is pushed upward into the air (Newton's 3rd Law at work).

NEWTON'S 3RD LAW

For every action, there is an equal and opposite reaction.

Action Force:

Gas pushes down out of the rocket

Reaction Force:

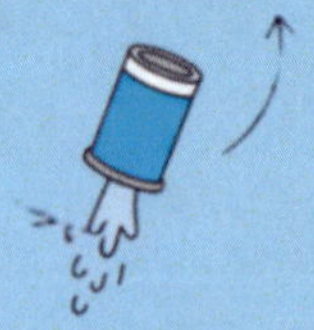

Rocket pushed up into the air

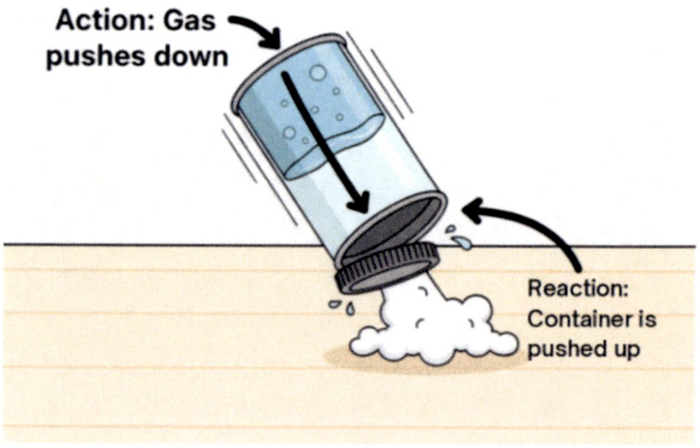

Take It Further with Your Fizz Rocket!

Want to keep the fun going? Inside the Secret Lab Club, I'll show you how to turn this simple launch into a full rocket mission — with bigger blasts, clever tweaks, and creative STEM challenges using your fizz rocket gear!

Next-Level Ideas Might Include:

Try a baking soda + vinegar rocket version for a louder, faster launch.

Discover ways to change the reaction speed using warm or cold water.

Explore rocket design ideas like fins, cones, and launchpads.

Simple tips for measuring flight height and launch distance.

www.thecrazyscientistlab.com.au

Predict what happens if you use different amounts of baking soda or vinegar. Test and record your results!

Use a phone timer to measure flight time and speed. Can you graph your results?

Build a stabiliser or launchpad so your rocket always takes off straight.

Decorate your rocket with fins, flames, and colours — make it look like a real space launch!

Launch three times and calculate the average height. How does it change with different amounts of fuel?

Gadget Wall Malfunction!

"WARNING! WARNING! Unauthorised Gadget Activation!"

Professor Picklebottom dangles upside down from a robotic grabber arm.

Dr Puddledrip sighs. "This is why we secure the Whiffinator before experiment days start."

A swirl of green mist escapes the Portal Reactor. The Gadget Wall hums like it's thinking.

"Junior scientists, brace yourselves. We're entering Unstable Portal Phase Two. Only fresh experiments can restore balance!"

Which Gadget Caused the Chaos?

Look at the Gadget Wall...

Which gizmo do YOU think triggered the malfunction? Match the picture to its name!

□ Shrinkascope □ Whiffinator 3000 □ Slimometer □ The Soapinator

Mini Maze: "Help the Portal Stabiliser!"

The Portal Reactor is wobbly!

Help Dr Puddledrip guide the stabiliser beam to the control panel.

Mission 05

High-Five : Swamp Monster Edition

In this crazy experiment, we're unleashing the swamp beasts! Three colourful monster hands will rise from their bubbling bogs... but which one will be first to high-five the sky? It's a race of science and slime!

Portal Progress

Before we start – which swamp beast will rise the fastest?

- **Frostbite Fingers (cold vinegar)**
- **Murky Marvin (room temperature)**
- **Lava Larry (warm vinegar)**

Make your guess before the race begins!

What You Need

- 3 × plastic bottles (250–500 mL)
- 3 × disposable gloves (different colours for monster hands)
- 3 × rubber bands
- Baking soda (1 tsp per glove)
- Vinegar (¼ cup / 60 mL per bottle)
- Ice + bowl (for Frostbite Fingers)
- Warm water bath (for Lava Larry)
- Funnel
- Measuring cups & spoons
- Stopwatch or phone timer
- Marker pens to draw monster faces

Step-by-Step:

1. Prepare the Swamp:

- Add 60 mL of vinegar to each bottle. Chill one in ice, keep one at room temperature, warm one in a water bath.

2. Create the Beasts:

- Fill each finger of the glove with 1 tsp of baking soda.
- Draw scary monster faces on each glove — claws, fangs, warts... the works!

3. Seal the Swamp:

- Attach each glove to a bottle without spilling the powder. Use rubber bands to seal them tight.

4. The Monster Awakens:

- On "GO," lift up the fingers of the gloves so the baking soda drops in. Stand back!

5. Race to the Sky:

- Watch as the monster's hands inflate and rise from the swampy bottles.
- Time each monster hand to see which wins the race.

Dr Puddledrip's Chemistry Breakdown

The chemical reaction here is the same one we've seen before:

Baking Soda (sodium bicarbonate) + Vinegar (acetic acid) → Carbon Dioxide Gas + Water + Salt

The fizzing bubbles of carbon dioxide fill the glove until it grows into a giant monster hand, ready to high-five the sky!

But this time, we're adding a twist: temperature.

- When the liquid is warmer, the particles move faster (see the diagram).
- Faster particles collide more often and with more energy, so the reaction speeds up.
- This means more gas is made in less time, and the glove inflates more quickly.

In cold liquids, particles move slowly → fewer collisions → slower reaction → the glove takes longer to inflate.

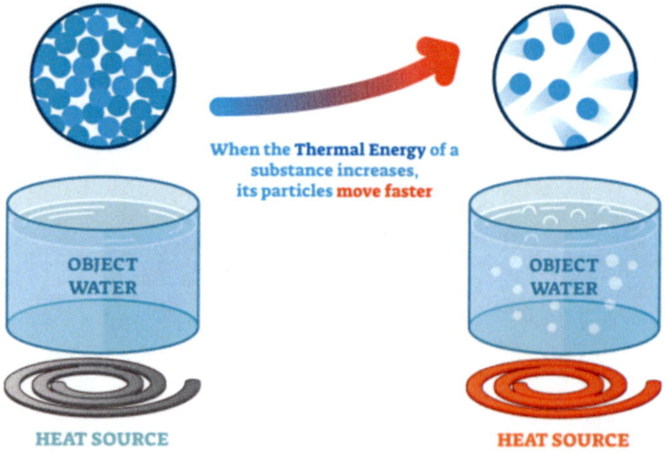

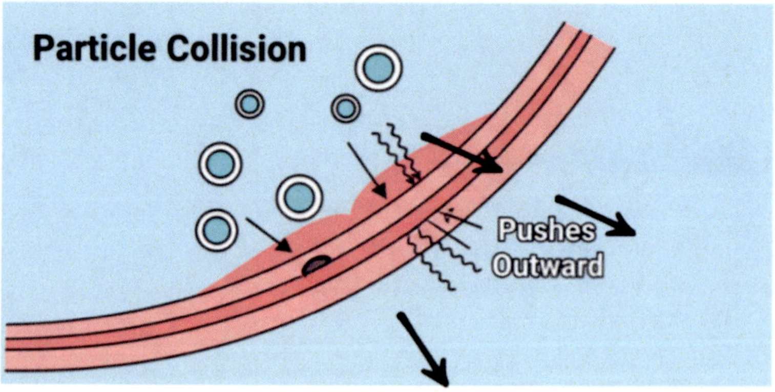

WOULD YOU RATHER...

Race a swamp monster that grows faster in warm water?
OR
Try to slow it down with ice-cold water so you have time to escape?
What do you think would happen — and why?

If you like this, then why not check out my Would You Rather book:

- **Crazy, silly Would You Rather questions**
- **Puzzles**
- **Experiments & More...**

Available at:

- www.thecrazyscientist.com.au (under Book Series)
- Amazon

Test different temperatures: cold, room temperature, warm, and hot (under 40°C). Graph glove-inflation time vs. temperature. What pattern do you see?

Use slow-motion video to capture the exact moment the glove starts to rise!

Build a cardboard or LEGO "swamp arena" so each glove inflates inside its own mini monster habitat.

Award prizes for "Scariest Glove," "Silliest Monster," and "Fastest Hands in the Swamp."

Run three races at each temperature.
Calculate the average inflation time for each temperature.

Mission 06

Volcano of Snot (Foaming Volcano)

Get ready for the grossest, gooiest eruption ever! In this experiment, we're building a slime-spewing volcano that blasts out stretchy, snot-like foam. It's messy, it's weird, and it's powered by a secret helper hiding in yeast that makes the reaction happen super-fast.

Portal Progress

Before we start — what do YOU think will change the eruption the most?

- **More yeast to speed up the reaction?**
- **A thicker slime mix for extra-stringy snot?**
- **Using warm vs cold water to wake up the yeast faster?**

Make your prediction — then let's see if you're right when the snot explosion begins!

What You Need

- A small bottle or tall beaker (your "volcano")
- 3% hydrogen peroxide (drugstore strength) — ½ cup (120 mL)
- Active dry yeast — 1 sachet (7 g)
- Warm water (35–40 °C) — ¼ cup (60 mL) to activate yeast
- Dish soap — 1–2 tablespoons
- Green/yellow food colouring (or two colours swirled for "pus effect")
- Thickener (choose one, a pinch to ¼ tsp):
  - Xanthan gum or guar gum (food-safe; gives snot stretch)
- Tray to catch the eruption, funnel, spoon/cup

Step-by-Step:

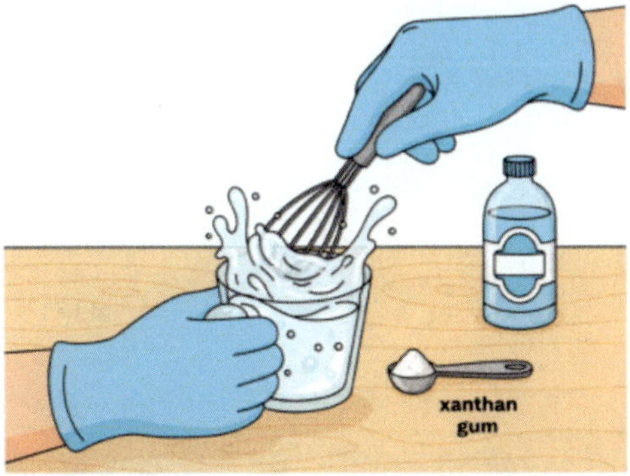

1. Prime the Snot:

- In a small cup, whisk ¼ tsp xanthan or guar gum into 1–2 Tbsp of dish soap until smooth.

(If you're only using a pinch, the foam will be less stringy and more bubbly — both versions work.)

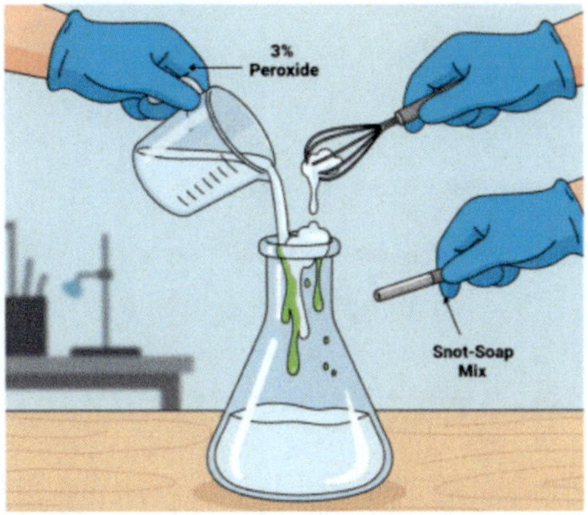

2. Load the Volcano:

- In the volcano bottle, add 120 mL of 3% hydrogen peroxide, your snot-soap mix, and a few drops of food colouring.

(streak some down the side for gross drips).

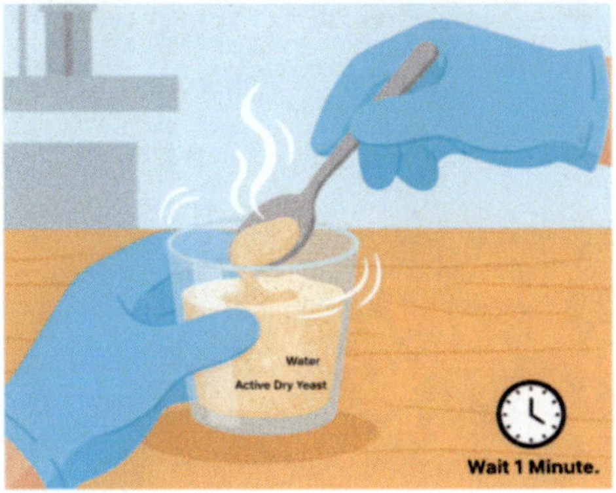

3. Wake the Yeast:

- Wake the Yeast: In another cup, mix the yeast with warm water. Wait 1 minute until the mixture becomes creamy and bubbly.

4. Erupt!

- Erupt! Quickly pour the yeast slurry into the volcano.
- Stand back and admire the thick, stretchy snot-foam surging out.

Dr Puddledrip's Chemistry Breakdown

When we mixed everything together, we made a chemical reaction:

Hydrogen Peroxide → Water + Oxygen Gas

This reaction is called a decomposition reaction. A larger chemical is broken down into smaller chemicals using a catalyst found in yeast.

What's a Catalyst?

A catalyst is a chemical that helps a reaction go faster, but it doesn't get used up — so it can keep working again and again!

In our volcano, the enzyme catalase inside the yeast is the catalyst. It quickly breaks the hydrogen peroxide into water and oxygen gas.

Why Do We Get Foam?

The oxygen gas bubbles up as the reaction happens.

The dish soap traps the gas inside lots of tiny soap bubbles.

The result? Huge amounts of foam come rushing out... and adding the thickener makes it look like stretchy, slimy snot!

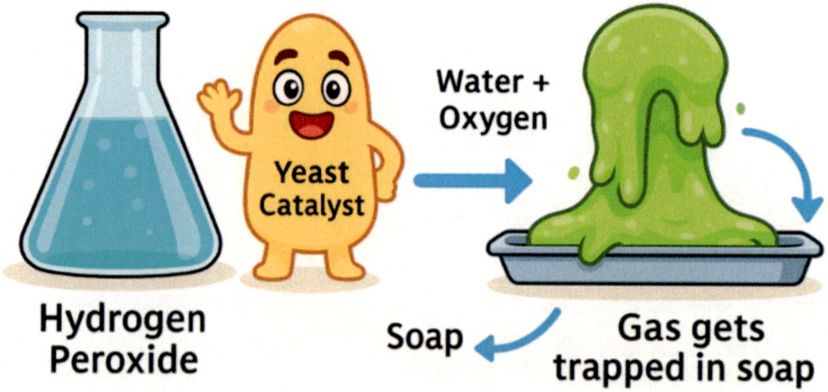

S

Test yeast vs no yeast to see if the reaction happens at all without the catalyst.

T

Build a DIY bubble counter with a clear tube or bottle cap so you can count how many gas bubbles are made before the eruption.

E

Design a "Volcano Containment Zone" using cardboard walls, so the slime doesn't escape your tray.

A

Give your volcano a gross monster face with fangs and eyes so it looks like it's spewing snot from its mouth.

M

Measure the time to first foam and the maximum foam height.

Mission 07

Soapzilla: Mutant Foam Beast Attack!

Run for your lives! The evil Professor Sudsy has unleashed Soapzilla, the monster that grows bigger and bigger every time it gets zapped in the microwave! Will science save the day — or will the entire kitchen be covered in fluffy soap foam?

Portal Progress

Before We Start – Make Your Predictions

Which Soapzilla will grow the most?

- **Baby Soapzilla (30 seconds)**
- **Teen Soapzilla (60 seconds)**
- **Mega Soapzilla (90 seconds)**

Write down your guess before you awaken the beast!

What You Need

- 3 identical pieces of Ivory Soap (cut the same size)
- Microwave-safe plate
- Microwave
- Timer or stopwatch
- Ruler or tape measure
- Safety gloves and an adult helper (the soap and plate get very hot!).

Step-by-Step:

1. Prepare the Arena:

- Cut the bar of Ivory soap into three equal pieces.
- Place the first piece on a microwave-safe plate in the centre of the microwave.

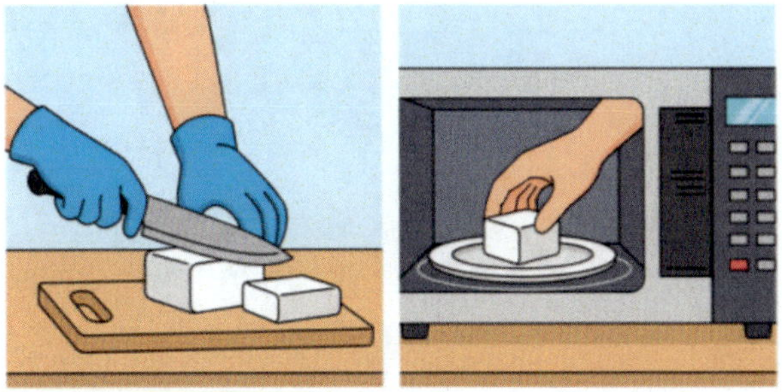

2. Zap Round 1: Baby Soapzilla

- Microwave for 30 seconds on HIGH.
- Watch through the window as Soapzilla starts to grow!
- Let it cool for at least 1 minute before touching.
- Measure the height, width, and length with a ruler.

3. Zap Round 2: Teen Soapzilla:

- Repeat the process with the second piece for 60 seconds.
- Let it cool, then measure its size.

4. Zap Round 3: Mega Soapzilla:

- Heat the final piece for 90 seconds.
- Again, cool before measuring the monster!

5. Crown the Winner:

- Compare all three measurements to see which Soapzilla grew the biggest.
- Record results in a simple table or bar graph.

Professor Pickelbottom's Chemistry Breakdown

When you put Ivory soap in the microwave, it doesn't just sit there – it mutates into a giant foam monster! But why?

The Secret Inside Ivory Soap

Ivory soap is whipped during manufacturing, so it's full of tiny air bubbles and a little bit of water trapped inside. That's why Ivory floats in water – it's less dense because of all that air.

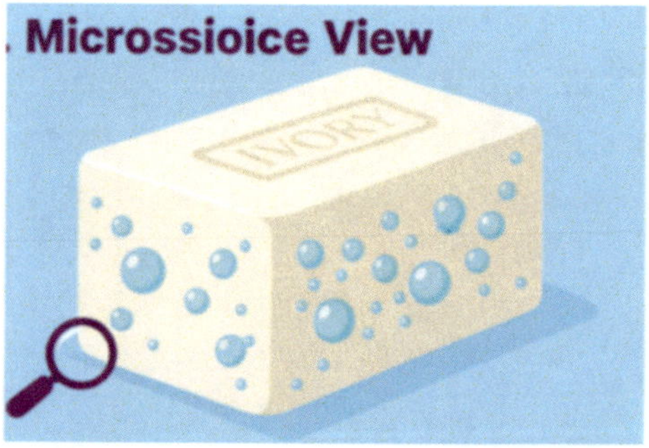

What Happens in the Microwave

Heat Attack!

The microwave heats the water molecules trapped in the soap, which makes them move faster and turn into steam (water vapour) as it gets hotter. The air bubbles inside the soap heat up too and also move faster, causing them to expand (taking up more space).

Soapzilla Transformation

Soapzilla Transformation: Expanding air + steam push against the soft soap walls. The soap stretches and puffs up like a monster, growing bigger and bigger!

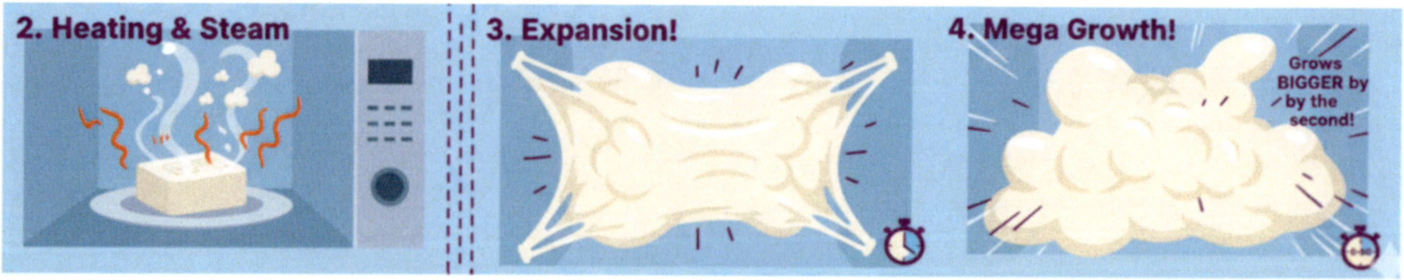

- **Physical change:** The soap changes shape and size but it's still soap — no new substances are made.
- **Gas expansion:** Hot gases (air + steam) have faster-moving particles that spread out more than cold gases.
- That's why longer microwave time = bigger Soapzilla (until all the water is gone).

Fun Fact: Why Only Ivory?

Other soaps don't expand as much because they're denser and don't have as many air bubbles trapped inside. No bubbles = no monster.

S — Compare Ivory soap to another soap brand. Does it still puff up?"

T — Use a time-lapse app or motion sensor to track how quickly Soapzilla grows.

E — Build a Soapzilla arena (cardboard city) for the monster to crush as it grows.

A — Create Soapzilla movie posters with dramatic titles like Revenge of the Suds!

M — Measure Soapzilla's length × width × height to calculate the volume before and after, then work out the percentage growth.

Radiation Attack: The Marshmallow Mutant Uprising!

Warning! Marshmallow mutants detected. Under microwave radiation, they swell into puffy beasts—then collapse back into gooey squish balls as they cool. Can our brave scientists control the uprising using the power of temperature?

Portal Progress

Before We Start – Make Your Predictions

Which one will grow the biggest and stay puffiest?

- **Freezezilla (from the freezer)**
- **Room Ruler (from room temp)**
- **Warm Warlord (gently pre-warmed*)**

What You Need

- 3 marshmallows (same brand/size)
- Freezer (for Freezilla, ~30 minutes)
- A warm spot (for Warm Warlord, ~5–10 minutes)
- Microwave + microwave-safe plate
- Ruler or tape measure
- Timer/stopwatch
- Marker (to label monsters)
- Optional: food pens for drawing faces

Safety

- Hot melted sugar can burn. Watch through the microwave window and let the marshmallows cool for 1–2 minutes before handling. Place a paper towel or microwave-safe mat on the plate to protect it, and clean the microwave afterwards.

Step-by-Step:

1. Name your monsters and label them:

- Freezezilla, Room Ruler, Warm Warlord.

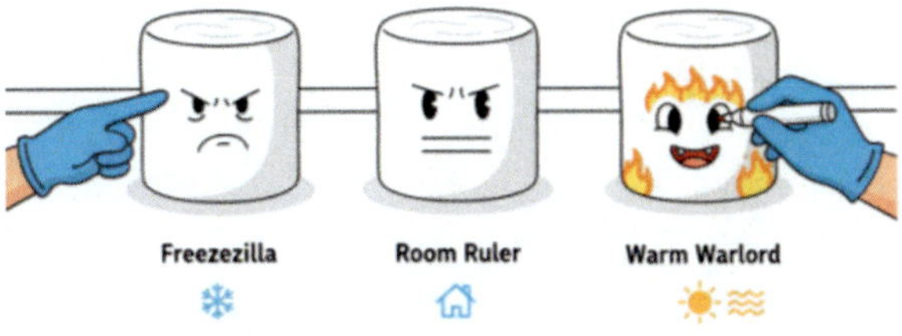

2. Prep the mutants

- Put Freezezilla in the freezer for 30 min.
- Leave Room Ruler on the bench.
- Place Warm Warlord near a warm mug/sunny spot for ~5–10 min.

3. Microwave Trial (each monster separately)

- Place the marshmallow in the centre of the plate.
- Microwave 30 seconds on HIGH. Watch the growth!
- Remove plate carefully and cool 1–2 minutes.

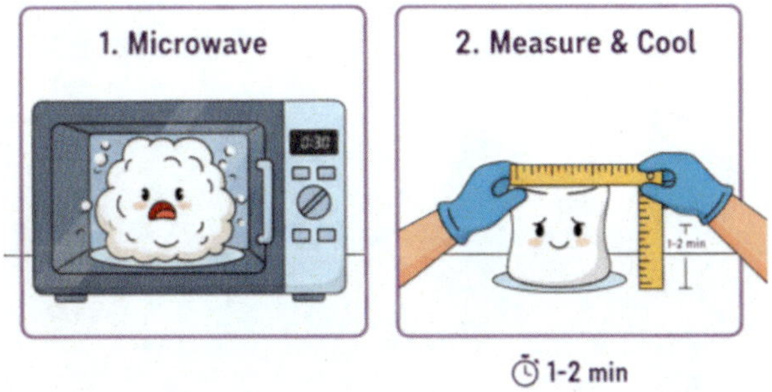

4. Measure the mayhem

- Record maximum height (or longest dimension) right after heating.
- After cooling 2 minutes, measure final size.
- Squish test (texture): gently press with a fingertip or toothpick and rate 1–5 (1 = brick, 5 = super squish).

5. Repeat

- For all three mutants.
  Crown the champion: Biggest puff, best "staying-puff," and squishiest.

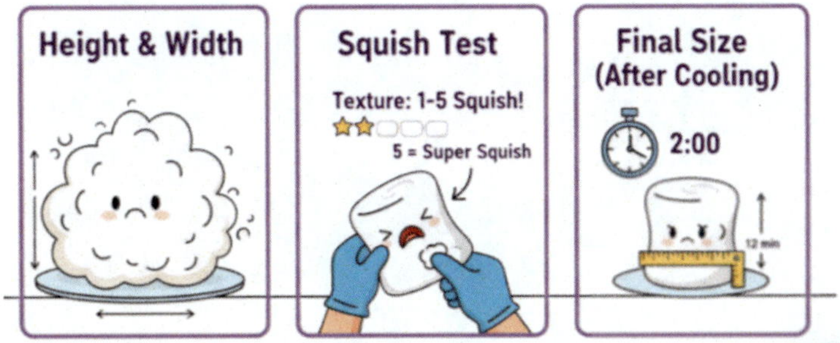

Marshmallow Monsters Data Table

Monster	Max Size After Heating (cm)		Size After 2 after 2 min Col (1–5)		Squisth (burst? sticky (1–?)	
Freezezilla	(~0–5 °C)					
Room Ruler	~20–22 °C					
Warm Warlord	~30–35 °C					

The Science of Marshmallow Monsters!

1. Inside the Monster

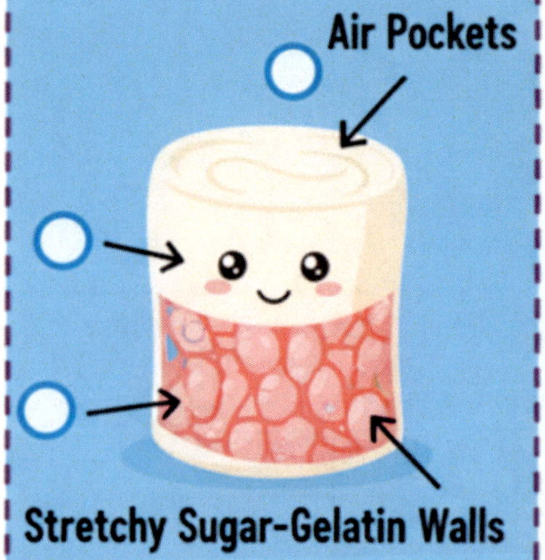

2. Microwave Attack!

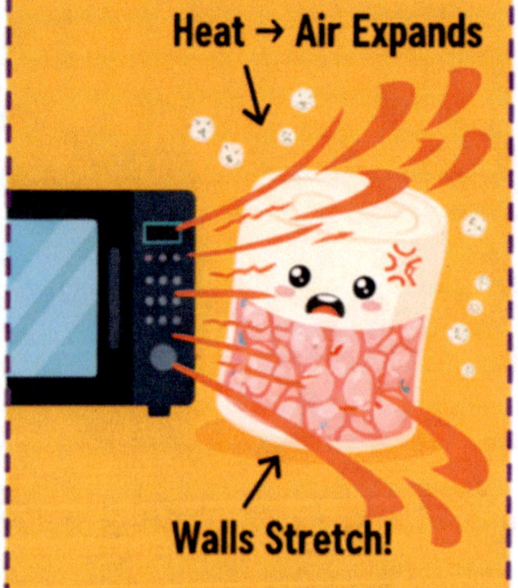

3. Cooling Down

Professor Pickelbottomp's Chemistry Breakdown

Marshmallows are foams: zillions of tiny air bubbles trapped in a stretchy sugar-gelatin network.

- **Heat** → air expands, and the gelatin softens → marshmallow puffs up.
- **Cool** → air contracts and the sugar network firms → marshmallow shrinks and can become chewier/harder.

Starting temperature changes the battle:

- Freezezilla (cold start): Air is dense,, and gelatin is stiff → slower expansion; cools fast → strong shrink-back, often firmer.
- Room Ruler: Balanced expansion and texture.
- Warm Warlord (warm start): Air expands quickly and gelatin is already flexible → biggest puff and softer texture (but may collapse more as it cools).

Key idea: Same marshmallow, different particle speeds. Warmer start = faster moving air molecules → more expansion during radiation attack.

Try wrapping identical marshmallows in different "insulators": a paper towel, or Bubble wrap.

Build a simple "puff gauge": a paper ruler taped upright behind the plate so you can read max height through the door.

Design a Mutant Containment Ring (cardboard collar) with height markings to capture maximum puff without sticking to the oven.

Monsterify each marshmallow with food pens; award "Scariest Puff," "Best Collapse," "Saddest Goo."

Calculate % expansion = (Max Size - Start Size) ÷ Start Size × 100

Mission 09

Lava Lamp Lunacy (gas + liquid separation)!

Dr Puddledrip has done it again! She's unleashed a glowing army of lava blobs that float, sink, and dance like crazy under the power of fizzing gas bubbles. But what mysterious science makes the blobs rise and fall in this wild lava lamp of lunacy?

Portal Progress

Before We Start – Make Your Predictions

Which will make the blobs move faster?

- **More Gas Power (extra fizz tablets)**
- **Different Liquids (oil vs water ratios)**
- **Temperature (cold vs warm water)**

Write down your guess before creating chaos in a bottle!

What You Need

- A clear plastic bottle or tall glass
- Vegetable oil (~3/4 full)
- Water (~1/4 full)
- Fizzy tablet (e.g., Alka-Seltzer, vitamin C)
- Food coloring (a few drops)
- Optional: small torch or glow stick for lava lamp effect

Step-by-Step:

1. **Layer It Up:**
   - Fill the bottle 3/4 full with oil, then add water until nearly full. Watch the layers form!

2. **Add Some Colour:**
   - Drop in 1–3 drops of food colouring.
   - They'll pass through the oil and colour the water layer below.

3. **Launch the Bubbles:**
   - Break the fizzy tablet into small pieces.
   - Drop one in and watch the bubbles race upwards, carrying blobs of coloured water into the oil.

4. **Glow It Up:**
   - Shine a torch from below or drop in a glow stick for full lava-lamp lunacy.

5. **Repeat the Chaos:**
   - Add more tablet pieces whenever the action starts to slow down.

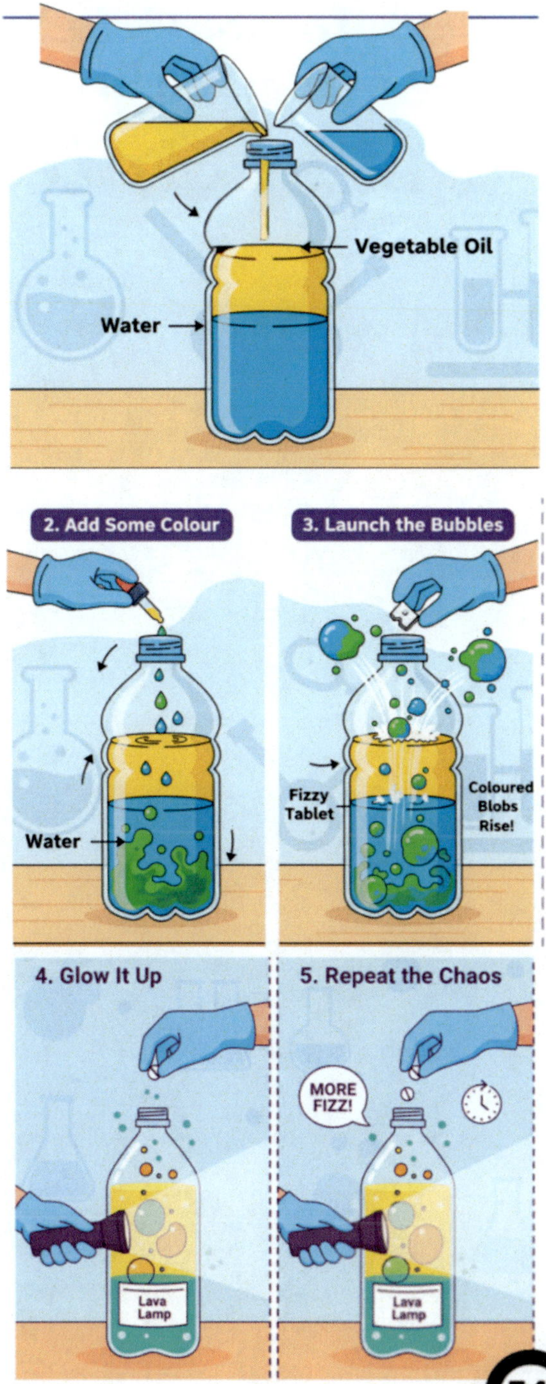

Professor Pickelbottom's Science Breakdown:

Lava Lamp Lunacy!

The Layer Trick

Water sinks and oil floats. They don't mix well because they have different densities!

The gas carries some of the coloured water upwards with it.

Then the gas escapes into the air, and the blob sinks back down to the bottom.
This keeps happening
until the chemical reaction stops making CO_2 gas

The Bubble Boost

Tablet + water produce CO_2 gas

DID YOU KNOW?

Real lava lamps use heat instead of fizzing gas! The heat changes the density of the wax blobs, so they rise and fall too.

Take It Further with Your Lava Lamp Lunacy!

““Ready to turn your Lava Lamp Lunacy into a real science mission? Test, tweak, and compare your results to discover what makes the craziest bubbling blobs!”

Next-Level Experiments:

- Try different water temperatures: Do warm blobs rise faster than cold ones?
- Change the amount of fizzy tablet: Does more gas mean more chaos?
- Experiment with oil types: Does thick cooking oil vs thin oil change blob speed?

www.thecrazyscientistlab.com.au

You'll learn about:

✓ Variables and fair testing
✓ Making predictions and recording results
✓ How temperature, gas, and liquid layers change density

S	Try hot vs cold water → does temperature change how fast the blobs rise?
T	Film in slow motion → count how fast the bubbles rise, frame by frame.
E	Build a bubble maze inside the bottle using straws or small plastic barriers.
A	Add two colours to the water → create rainbow blobs or alien invasion scenes.
M	Time how long each blob takes to rise and sink → graph the results.

Mission 10

Mini Firefighter: The CO_2 Extinguisher Challenge!!

The Candle of Chaos is back! But this time, it's facing two firefighter squads:

- Team Polar Rescue armed with icy vinegar
- Team Solar Blasters with sun-warmed vinegar
- Which squad will blast out the flames the fastest?"

Portal Progress

Before We Start – Make Your Predictions

Which will put out the flame quicker?

- **Icy Vinegar (slower reaction?)**
- **Room Temp Vinegar (medium reaction?)**
- **Sun-Warmed Vinegar (super-fast fizz?)**

Write down your guess before the battle begins!

What You Need

- 3 tea-light candles on a tray
- 3 small clear cups or beakers
- Baking soda (2 Tbsp per cup)
- Vinegar: one chilled, one room temp, one warm (not hot — ~40°C)
- Matches/lighter (adult job)
- Thermometer (optional, for older kids)
- Stopwatch or phone timer

Step-by-Step:

1. Set the Scene:

- Place three candles on the tray and label them:

  Cold, Room Temp and Warm.

2. Light the Flames:

- Adults light all three candles at the same time.
- Only adults should handle flames.

3. Charge the Extinguishers:

- Add baking soda to each cup.
- Pour the different vinegar temps into each labelled cup.

4. Tilt and Extinguish:

- One at a time, tip the cup so only CO_2 gas (not liquid) flows gently over the candle flame.

5. Time It:

- Start the stopwatch when you pour. Stop when the flame goes out.

Compare Results:

Which temperature put the flame out the fastest?

The Science Bit: Solar-Powered Fire Extinguisher

Vinegar (acid) + **Baking soda (base)** = **CO_2 + Water + Salt**

Fizz Power!

What's Happening in the Cup?

- When vinegar (acid) meets baking soda (base), they react to form carbon dioxide gas (CO_2), water, and a type of salt.
- The fizzing you see? That's **CO_2** gas bubbles racing out of the liquid.

Reaction Race

Why Temperature Matters:

- Warmer vinegar makes particles move faster. Faster-moving particles collide more often and with more energy.
- More collisions = quicker chemical reaction = **CO_2** produced faster.
- Faster **CO_2** production creates a thicker, denser **"gas blanket"** over the fiame.

Carbon dioxide gas falls downwards and pushes oxygen from the flame.

No oxygen = No fire!

S	Test what happens if you remove each ingredient of the Fire Triangle: heat, fuel, or oxygen. Example: Cover a candle with a jar (no oxygen).
T	Compare how fast the reaction happens with warm vinegar vs cold vinegar.
E	Build a cardboard or paper 'extinguisher nozzle' to guide the CO_2 gas toward the flame (only adults handle flames!).
A	Draw the Fire Triangle, label the three elements, and show how CO_2 removes the oxygen.
M	Use a stopwatch to measure how long it takes the flame to go out using warm vinegar vs cold vinegar.

Fizz-Powered Submarine Showdown!

Oh no! The Mini-Sub Explorer is trapped deep under the sea with a belly full of fizz fuel. Only you can launch it back to the surface! But beware... too much fizz and it might burst early. Too little and it'll stay on the ocean floor forever. Can your crew save the sub in time?

Portal Progress

Before We Start – Make Your Predictions

- **Which will make the sub rise the fastest?**
- **More fizz tablet pieces or fewer?**
- **Does warm water make the sub rise faster than cold water?**
- **Does a bigger vent hole make the sub sink more quickly?**

What You Need

- One small plastic bottle with a screw-on lid (travel bottle, mini soda bottle, or film canister)
- Fizzing tablets (effervescent vitamins or DIY citric acid + sodium bicarbonate)
- Plasticine or Blu Tack for adding weight
- Water tub or large container filled with water
- Thumbtack or safety pin (for making the vent hole – adult help recommended)
- Measuring cup or small jug

Step-by-Step:

1. Build the Sub

- Use a small plastic container (e.g., travel shampoo bottle, mini soda bottle, or film canister).
- Poke a tiny vent hole in the lid with a thumbtack. (Adult help recommended.)

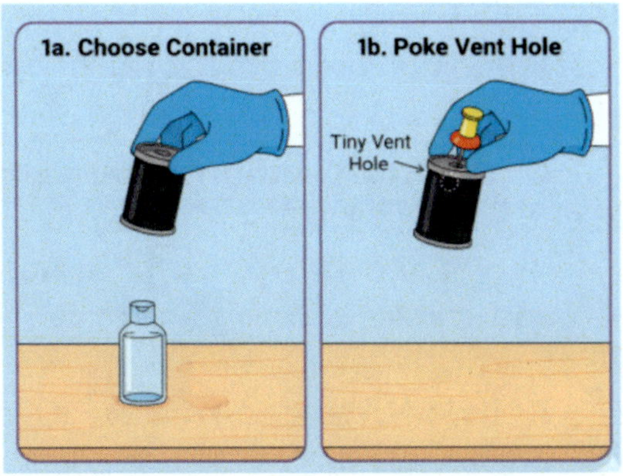

2. Ballast Testing (No Fizz Yet!)

- Add water until the sub just barely sinks.
- If it floats too much, add small blobs of plasticine or Blu-Tack on the bottom for weight.
- The goal: a perfect submarine should sink when full, but rise when the fizz pushes water out.

3. Fizz Power!

- Empty out the water, then refill to the same level.
- Drop in half a fizzing tablet (citric acid + sodium bicarbonate).
- Seal the lid tightly.

4. Launch the Sub

- Place it gently in a large tub of water.
- Watch as bubbles push water out, making the sub lighter so it rises.

5. Dive Again

- As gas escapes through the vent hole, water leaks back in. The sub sinks again!

Safety: An adult should help make the vent hole and check that the lid can't pop off.

Dr Puddledrip's Chemistry Breakdown

When you drop the fizz tablet into the water inside the mini-sub, it starts a chemical reaction:

Citric acid + sodium bicarbonate → carbon dioxide gas (CO_2) + water + a type of salt

At first, the mini-sub is full of water, so it is heavy. The downward pull of gravity is stronger than the upward buoyant force from the water, so the sub sinks.

As the fizzy reaction makes CO_2 gas, the gas bubbles fill the mini-sub, pushing some of the water out. Now there's more gas and less water inside, so the mini-sub becomes less dense than the water, and it rises.

The upward buoyant force is now stronger than gravity, and the sub floats up.

When the sub reaches the top, the gas escapes and water rushes back in. The sub gets heavy again and sinks back down. This rise-and-fall cycle keeps going until the fizzing stops.

Fish Do It Too!

Fish use a special organ called a swim bladder. By changing how much gas is inside, they control whether they rise or sink in the water—just like our mini-sub!

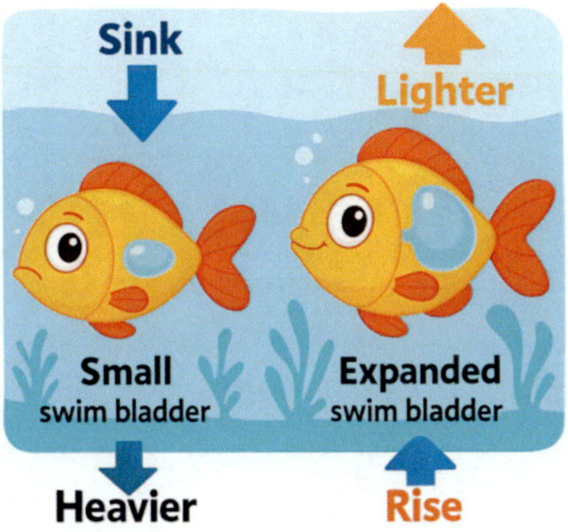

Add salt to the water and see how that affects the buoyancy. Do you need to add or remove more weight from the container?

Make a simple scale on the side of the container with tape or a marker. Track how deep the sub sinks each time as the fizz slows down.

Make two or three different subs with small changes — size, weight, hole size in the lid — and see which one gives the best rise-and-fall cycle.

Decorate the outside of the container to look like a real mini-sub or a deep-sea explorer. Give each one a fun name and design a mission badge!

Estimate the number of bubbles per second at the start vs near the end. Use this to predict when the sub will stop moving.

Portal Malfunction!

"WARNING! WARNING! SCALE SHIFT DETECTED!"
"ENVIRONMENT CLASSIFIED: ORGANIC. WET. VERY, VERY WET."

With a sudden SPLOOOORP! the world stretches, wobbles… and then shrinks around them.

The floor squishes. The air smells… suspiciously salty. Everything glows a warm, gooey green.

Dr Puddledrip blinks. "Oh no…" she whispers.

"Oh YES!" giggles Professor Picklebottom.

Before she can stop him—

SCHLOOOP! A giant booger boulder rolls past like an avalanche, splattering goo everywhere.
Dr Puddledrip leaps sideways, narrowly avoiding a sticky splat.

Behind her, Picklebottom is dangling upside-down, hopelessly stuck to a wall of nose hairs.

"Interesting!" he says cheerfully, still hanging.
"The nasal cavity is far more spacious than the textbook drawings!"

A low rumble begins… The entire tunnel shakes…
Dr Puddledrip freezes.

"Oh no, this looks suspiciously like the inside of a dinosaur nostril."

She points toward the dark tunnel and says, very slowly,
"A SNEEZE is building…"

The airflow shifts. A roaring wind gathers. Picklebottom's hair stands straight up.

"JUNIOR SCIENTISTS — BRACE YOURSELVES!" Puddledrip yells.

"If we don't escape before the sneeze fires, we'll be launched into low Earth orbit!"

She taps her holo-tablet, scanning the gooey tunnels.
"There! The mucus river leads straight to the Sinus Junction.
If we stabilise the air pressure, the Portal Reactor can pull us out!"

Picklebottom flaps helplessly.
"In that case… could someone please unstick me?!"

"At this size, one sneeze could launch us across the classroom!" cries Dr Puddledrip.
"Quick! Complete the mini-mission so we can escape before—"

A-A-AACHOOO!

Turn the page for Experiment #11 before the sneeze hits!

Mission 11

Super Soap Domes

Do you have the bubble-building skills of a master architect? Using the perfect bubble juice, see how many bubble domes you can stack one inside another before the whole bubble tower bursts!

Portal Progress

Before We Start – Make Your Predictions

- **How many domes will you stack before it pops?**
- **Will thicker bubble juice help you build more layers?**
- **Does resting the bubble mix longer make stronger domes?**

What You Need

- Mixing bowl
- Measuring cups and spoons
- Whisk or spoon (for gentle stirring)
- Flat tray or baking dish (to pour bubble solution into)
- Dishwashing solution (concentrated)
- Glycerin
- Drinking straws (for blowing bubbles)
- Stopwatch or timer (for extra challenges)

Step-by-Step:

1. Prepare the Table

- Pour a little bubble juice or water on the table.
- Spread it with your hand so the surface is wet (this stops the bubbles from popping too soon).

2. Dip and Blow

- Dip your straw into the bubble juice.
- Hold the straw so it just touches the wet table.
- Blow gently and steadily to make a big bubble dome on the table.

3. The Magic Trick

- Dip the straw back into the bubble juice so it's wet again.
- Gently push the wet straw through the bubble wall (it won't pop because it's wet!).

4. Blow More Domes Inside

- Blow softly to make a second bubble inside the first bubble.
- Keep repeating to see how many bubbles you can make inside the first one!

5. The Challenge

- Try for the bubble dome record—how many bubbles inside one bubble can you make before it pops?

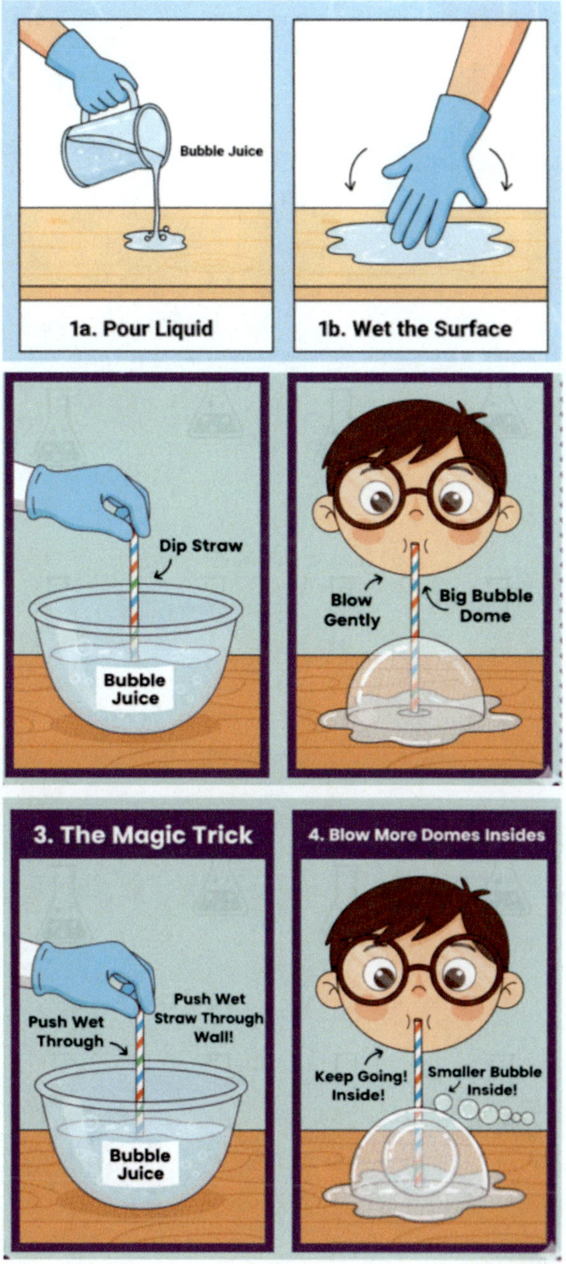

Professor Picklebottom's Chemistry Breakdown

The bubble juice is made of soap and water molecules. The water molecules are strongly attracted to each other, like people holding hands in a chain.

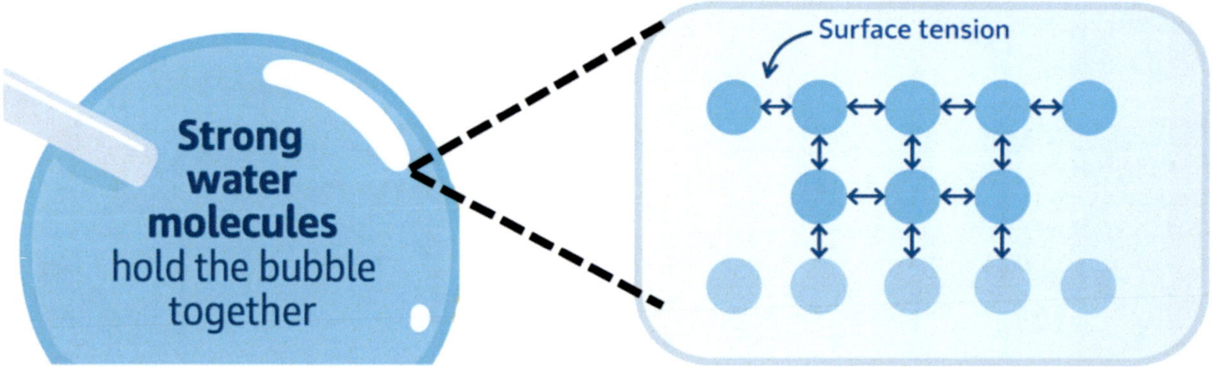

This attraction forms a stretchy "skin" on the surface of the bubble. This is called surface tension.

- When the straw is wet, it slides between the molecules without breaking their 'hand-hold,' so the bubble stays intact.
- When the straw is dry, it snaps the 'hand-hold' between molecules—just like breaking a chain of people—so the bubble pops!

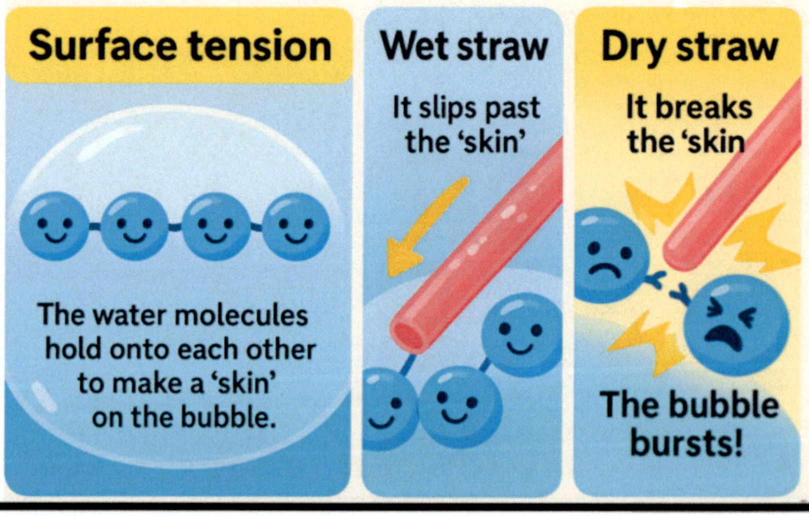

The Perfect Bubble Dome Juice

You'll Need:

300 ml water + **dishwashing liquid** (non-toxic) + **1 tbsp glycerin or glucose syrup** (for super strong bubbles!)

Steps to Make It:

1. **Mix Slowly**
   Pour water and dish soap together gently—don't make foam!

> Resting the mixture makes stronger bubbles!

2. **Add the Secret Ingredient**
   Stir in glycerin or glucose syrup. This makes the bubbles last longer.

3. **Let It Rest**
   Leave the mix fof at least 1 hour. Overnight is even better!

4. **Ready to Use**

> Pour into a flat tray for building giant bubble domes.

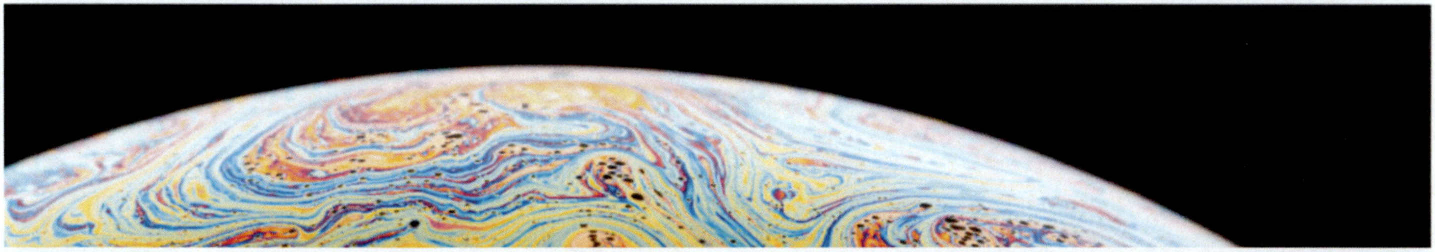

S — Test how temperature affects bubble strength: try cold water vs warm water in the mixture. Which temperature makes the bubbles last longer?

T — Design a simple bubble strength tester using paper clips or straws. Gently lower a weighted straw onto the bubble dome to see how much weight it can hold before popping.

E — Make a cube-shaped frame so you can create a square bubble.

A — Add food colouring to the bubble mix for rainbow bubbles. Take close-up photos to see the swirling patterns.

M — Measure the diameter of the biggest bubble dome you can make — use a ruler or tape measure.

Mission 12

The Big Fizz-Off!

Professor Picklebottom has gathered the fiercest fizzy fuels in the land: Lemon Lightning, Orange Overdrive, Cola Chaos, and Citric Cyclone. Only ONE can win the ultimate title of Fizz-Off Champion! Which fuel will inflate its balloon the fastest? Which will make the biggest balloon? Time to find out in... The Big Fizz-Off!

Portal Progress

Before We Start – Make Your Predictions

- **Which liquid will be the fastest fizzer?**
- **Will the strongest acid make the biggest balloon?**
- **Will the fizziest fuel change colour the most with the pH indicator?**

What You Need

- Four 600 mL plastic bottles
- Four balloons
- Baking soda (3 tsp per bottle)
- ½ cup each of:
    - lemon juice, orange juice, cola
    - citric acid solution (½ tsp citric acid in water)
- Funnel and measuring spoons
- Red cabbage indicator (optional, for pH colour test)
- Stopwatch or timer
- Marker pens for labelling bottles

Step-by-Step:

1. Name the Racers:

- Label each bottle with its fuel name (Lemon Lightning, Orange Overdrive, etc.).

2. Pour the Fuel:

- Add ½ cup of each liquid into its matching bottle.

3. Load the Booster:

- Use a funnel to put 3 tsp of baking soda into each balloon. Keep the balloons upright, so the powder stays inside.

4. Ready... Set... GO!

- At the same time, tip each balloon upright so the baking soda drops into the bottle.

Start the Timer: see which balloon inflates the fastest, and which is the biggest after 60 seconds.

5. Bonus Twist:

- Add red cabbage juice to each fuel before the race to see pH colours change as the reaction happens!

Professor Picklebottom's Chemistry Breakdown

When you mix baking soda (a base) with an acid (like vinegar, lemon juice, or orange juice), they react in what's called a neutralisation reaction.

This reaction makes three things:

1. **Carbon dioxide gas (CO_2)** → inflates the balloon and makes all the fizz.
2. **Water** → stays in the liquid.
3. **A type of salt** → is harmless and left behind in the mixture.

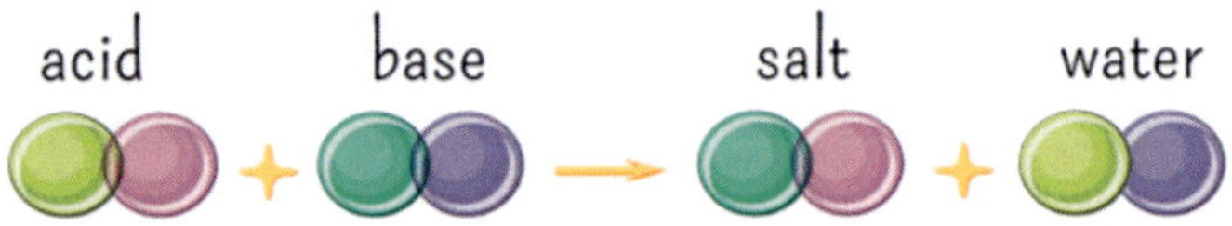

Why Some Acids Make Bigger, Faster Balloons?

Stronger acids - release more hydrogen particles, and they release them faster, so more gas is made in less time. Bigger, faster balloons!

This means the neutralisation reaction happens fast, producing more CO_2 gas in less time → balloon inflates faster and bigger.

Weaker acids release hydrogen ions slowly or have fewer of them.
The reaction takes longer and makes less CO_2 gas → balloon inflates slower and smaller.

Stronger acids release more positive particles that react with the base, causing gas to be produced.

S	Rank each fuel by speed, balloon size, and pH colour → compare the results.
T	Use slow-motion video to see which balloon expands fastest.
E	Design a tipping frame so all balloons release baking soda at the same time → fair race!
A	Decorate each bottle like a racing car with faces, names, and flames.
M	Measure the time it takes each balloon to reach a certain size → turn it into a speed graph.

Steam-Powered Snack Attack!

Kernel Commanders, to your stations! Each tiny kernel is a pressure capsule. Heat turns the water inside into steam. Pressure builds... then KABOOM!—the shell bursts and the starch inflates into popcorn.

Portal Progress

Which kernels will pop fastest and make the fluffiest explosion?

- **Cold kernels (fridge)**
- **Room-temp kernels**
- **Warm kernels (gently pre-warmed in a sealed bag next to a warm mug)**

Write down your guess before you begin the first launch!

What You Need

- Three small portions of plain popcorn kernels (2 Tbsp each)
- Paper lunch bags or a lidded saucepan and a little cooking oil
- Oven mitts, a timer, and a heat-safe bowl
- Ruler or measuring cup for volume, and a notebook

Safety

Adult supervision. Hot steam burns; open bags/pots away from you.

Step-by-Step:

1. Prep squads:

- Label the bags “Cold”, “Room”, and “Warm”.
- Put 2 Tbsp of kernels in each.

2. Set temps:

Cold: refrigerate for 30 min.

Warm: rest the sealed bag beside a warm mug or heat pack for 5–10 min (not hot).

3. Launch:

- Microwave each bag separately on HIGH for 1½–2½ min. Stop 2–3 seconds after the pops slow down.
- Start the timer when you press start.

4. Measure the blast:

- Pour each batch into the same measuring jug.
- Record the popped volume (cups) and the unpopped kernels left.

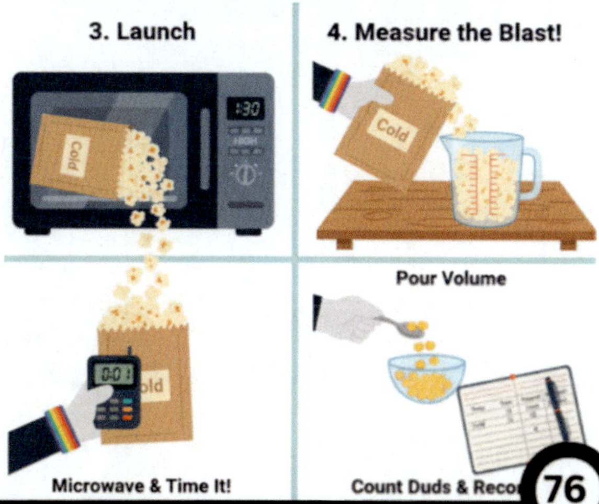

5. Rank the explosions:

- Fastest popping time? Biggest volume? Fewest duds?

(For stovetops: use the same amounts in a lidded pan with a teaspoon of oil. Shake gently as the popping starts. Measure as above.)

Professor Picklebottom's Chemistry Breakdown

What's Happening Inside the Kernel?

HEAT ATTACK!

- When we heat the popcorn kernel, the water inside turns into steam (like a tiny kettle!).
- This is a physical change, where water turns from a liquid into a gas when it gets more energy.

Pressure Build-Up!

- The steam can't escape because the kernel's hard shell is like a tiny pressure cooker.
- More steam = more pressure pushing outwards.

Kaboom Moment!

At about 180 °C (that's super hot!), the pressure gets so strong the shell bursts open – BOOM!

The hot starch and steam mixture inside puffs up and cools down super quickly, turning into the crunchy white foam we call popcorn!

Why Changing the Starting Temperature Matters

Heat Attack!
When we heat the popcorn kernel, the water inside turns into steam → Pressure builds faster → Popcorn pops quicker.

Cold kernels
take longer to heat up → **slower pressure build → delayed explosions.**

Too dry kernels
might not pop at all = **no water = no steam = no kaboom!**

Day 2 variable = moisture. Compare fresh kernels with kernels dried overnight on a tray, and kernels sealed with ½ tsp water to rehydrate for 24 h.

Use a sound meter app to log peak "POP!" loudness (dB).

Build a simple pop-containment tower (clear tube + paper funnel) and mark the height reached by the first popped flake.

Design movie-poster labels for each squad: "Chill Thrillers," "Room Rangers," "Heat Heroes."

Calculate popping efficiency = popped volume ÷ time (cups per minute). Graph starting temperature vs. time and efficiency.

Mission 14

Earth-in-a-Jar!

Now's your chance! Today, you'll explore what Earth looks like inside — without melting your shoes. Grab your goggles and get ready to pour, layer, and unzip the planet!

Portal Progress

We're going to use colourful liquids to model Earth's four main layers — the crust, mantle, outer core, and inner core. Each liquid will behave differently, just like real rock and metal do deep underground. By the end, you'll have your very own mini Earth — no volcano insurance required.

Before pouring, guess what will happen:

- **Which liquid will float to the top?**
- **Which will sink to the bottom?**
- **Will the layers stay separate, or mix like a cosmic milkshake?**

What You Need

- 1 clear plastic cup or jar (your "Earth")
- 1 small measuring cup or spoon
- Honey or golden syrup – Inner Core (heaviest)
- Dishwashing liquid – Outer Core
- Coloured water – Mantle
- Vegetable oil – Crust (lightest)
- Food colouring (optional)
- Stirring stick or spoon
- Ruler or marker (for layer lines)

Step-by-Step:

1. The Inner Core (honey or syrup – densest)

- Pour about 2 cm of honey or syrup into the bottom of your jar.
- This is the solid metal heart of Earth where it's hottest.

(**Tip:** it's sticky, so go slow!)

2. The Outer Core (dish soap)

- Slowly pour 3 cm of dish soap on top of the honey.
- This layer is made of liquid metal – it flows and spins to make Earth's magnetic field.

3. The Mantle (coloured water)

- Add 4 cm of coloured water.
- This is the thickest part of Earth – a slow-moving ocean of rock!
- Watch how the blue layer sits above the soap.

4. The Crust (oil – lightest)

- Finally, drizzle 1 cm of vegetable oil on top.
- That's your crust – thin, light, and floating just like the continents do.

Professor Picklebottom's

EARTH-SPLOSION Explanation!

How your jar shows the secret layers hiding inside our planet!

When Earth Was Born

It started as a giant ball of hot, melted rock and metal. Everything was mixed up — like space soup!

Gravity at Work!

Over time, gravity pulled the heaviest metals (like iron) down toward the middle. Lighter rocks floated up to form the top layers — just like your jar!

IN YOUR MODEL:

Layer	Material	Helght	Fun Fact
Crust	Vegetable oil	0.5 cm	Lightest– floats on top
Mantle	Coloured water	4 cm	Biggest laver – rock flows slowly
Outer Core	Dish soap	3 cm	Liquid metal makes magnetic field
Inner Core	Honey / syrup	1.5 cm	Solid metal heart hottest part of Earth

Did You Know?

If Earth were size of an an apple, the crust would be thinner than apples skin!

Take it Further with Exclusive Parent & Teacher Resources

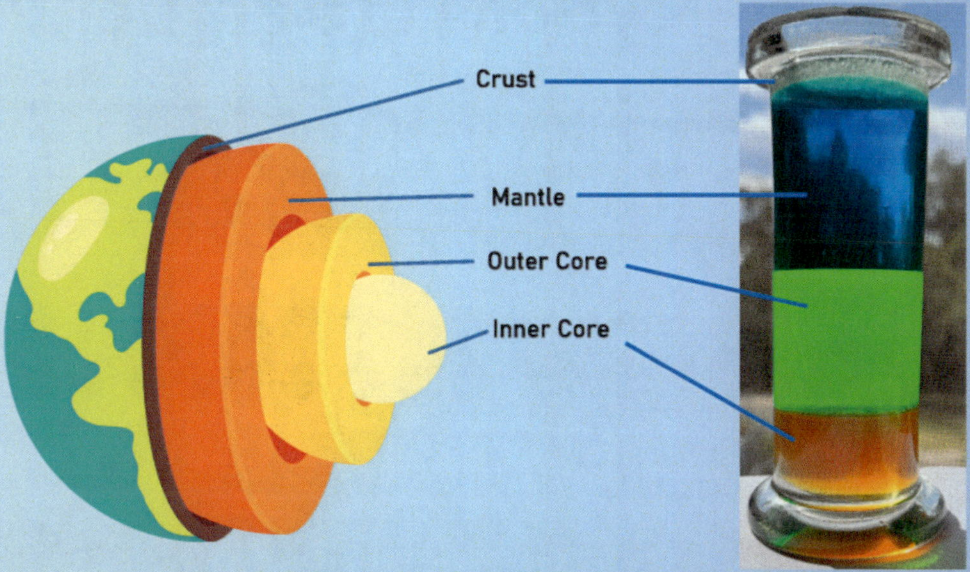

Make sure the thickness of your layers is in a similar ratio to the real Earth's layers. Learn how to calculate this in our Secret Lab membership, with extra worksheets and videos!

Test what happens when you swap liquids! Try honey vs. golden syrup, or warm vs. cold water. Which one makes the layers change faster or mix more?

Use a tablet or phone to take time-lapse photos of your jar. Do the layers stay separate, or do they slowly change?

Build a "Planet Pouring Station" — a cardboard or LEGO stand that keeps your jar steady so you can pour each layer perfectly.

Draw or paint a cross-section of Earth with faces and names for each layer (e.g., "Manny Mantle," "Corey Core"). Add speech bubbles showing what each layer might say!

Measure the total height of your jar.
Calculate what percentage each layer takes up.
Then draw a pie chart or bar graph to show your results.

Cyclone in a Bottle!

Why does the water spin into a swirling cyclone instead of just glugging straight down? What's really happening inside that whirlpool? Let's find out what happens when science takes a spin — using nothing more than water, two bottles, and a twist of the wrist!

Portal Progress

Before we start — what do YOU think will happen when you flip the bottles? Will the water glug slowly, gush out all at once, or spin into a tornado?

What You Need

- Two clear plastic bottles (1.25 L or 2 L PET)
- Bottle connector or strong tape
- Water (fill one bottle nearly to the top)
- Optional: glitter or food colouring

Try It Another Way!

Swap the water for liquids of different thicknesses — like juice, oil, or syrup — and see how your cyclone spins!

Need a connector?

Check them out in our online store.

Step-by-Step:

1. Fill It Up!

- Fill one bottle about halfway with water.
- Add a drop of food colouring or a pinch of glitter if you want extra "storm sparkle".

2. Connect the Bottles

- Place the washer over the mouth of the filled bottle, then screw on the connector.
- Attach the empty bottle to the other end.

If you don't have a connector, use strong tape to seal the bottles tightly.

3. Flip the Storm

- Hold the bottles with the full one on top.
- Flip them so the water-filled bottle is now upside down over the empty one.

4. Spin the Cyclone

- Give the top bottle a quick, smooth swirl in a circle. Watch as a swirling cyclone forms, acing the water down!

5. Time the Twister

- See how long it takes for all the water to drain. Try it again with a faster or slower swirl — does it change the speed of your cyclone?

What's Going On?

The key idea here is that air takes up space!

In the lower bottle, the air has nowhere to go — it's blocking the water from flowing down. The air in the bottom bottle pushes upward while gravity pulls the water in the top bottle downward. It's a traffic jam between air and water.

When you spin the bottle to make a vortex, the water swirls to the sides, leaving an open tunnel in the middle. This tunnel gives the trapped air a clear path to move up into the top bottle while the heavier water flows smoothly down into the bottom bottle.
It's all about air pressure and flow paths.

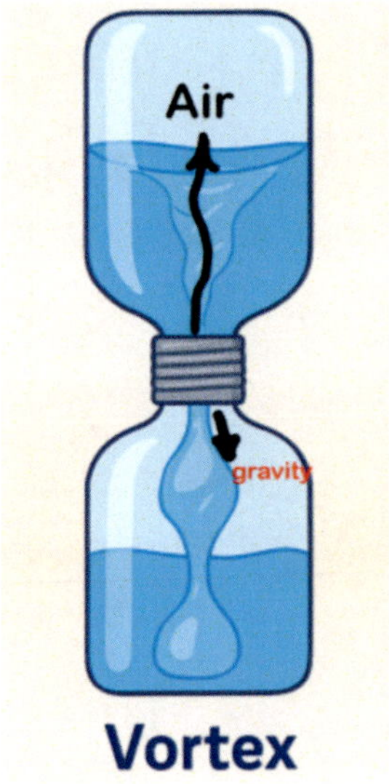

Two Ways Water Flows Between Bottles

Glug-Glug Flow
Water drops in bursts because air can only bubble up through tiny gaps. This makes the water "glug" and slows it down.

Vortex Flow
Spinning the bottle makes the water swirl to the sides, leaving a tunnel in the middle for air to rise. Water and air swap places smoothly, so it drains faster.

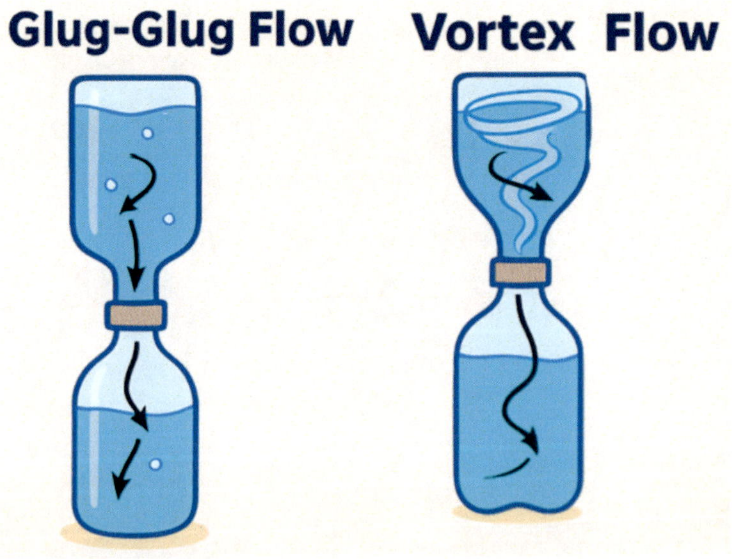

Try different amounts of water in the top bottle (full, half, quarter) and measure how long it takes to empty each time. Which amount drains fastest?

Use a phone or tablet to film the vortex in slow motion. Watch closely to see how the air tunnel forms and changes as the water drains.

Design and build your own connector for the bottles using cardboard, tape, or recycled materials. Test whether different connector shapes change the vortex speed.

Create a comic strip or animation of a "vortex adventure," giving the water and air characters personalities as they race to swap places.

Time how long the vortex takes to empty for different water levels or spin speeds. Graph your results to find the fastest combination.

Mission 16

Minion Levitation

Use the power of air to make your Minion float like magic!

Have you ever wondered how planes and birds stay up in the sky? It's all thanks to the power of moving air and changes in pressure. This invisible force can keep kites soaring, make a shower curtain mysteriously cling to you when the water's running — and even help a Minion balloon float in mid-air.

Portal Progress

Before we start — what do YOU think will happen?

Will your Minion balloon shoot straight up, hover in one spot, or wobble side to side?

What You Need

- One balloon (Minion-themed if you have one!)
- One hair dryer (set to cool or low heat)
- One small coin (10c or 20c) placed inside the balloon before inflating — this helps it stay steady
- Optional: markers or stickers to decorate your balloon
- **Parent or adult supervision** when using the hairdryer

Step-by-Step:

1. Load the Minion (adult help)

- An adult stretches the balloon's neck and slides a 10c or 20c coin inside.
- Gently shake or tilt the balloon so the coin rests near the knot — this helps the balloon stay steady in the airflow.

2. Inflate & decorate

- An adult inflates the balloon to a medium size and ties a knot.
- Optional: draw a Minion face with markers or add stickers.

3. Set the hairdryer

- Place the hair dryer upright on a stable table with the nozzle pointing straight up.
- Set it to Cool (best) or Low Warm, with Low or Medium airflow.
- Make sure the air intake is clear, and long hair or loose clothing are tied back.

4. Make it hover!

- Hold the balloon by the knot and place it about a hand's width above the nozzle.
- Turn the dryer on and slowly move the balloon up or down to find the balance point where it begins to float.

Nudge it left or right — or tilt the dryer slightly — to "steer" your Minion.

Try It Another Way

- **Coin or no coin?** Compare a balloon with the coin and without the coin. Which one hovers more steadily?
- **Balloon size test:** Try small, medium, and big balloons. Does the size change the "hover height"?
- **Height challenge:** Move the balloon higher or lower and time how long it can hover at each height.

Safety

Adult supervision required.
Use Cool or Low Warm airflow only.
Keep fingers, hair, and loose clothing away from the air intake.
Never aim the dryer at faces.

What's Going On?

The main idea here is air pressure — air always pushes from areas of high pressure towards areas of low pressure.

When air from the hair dryer blows around the balloon, it moves faster. According to the Bernoulli principle, faster-moving air has lower pressure.

So the air just around the balloon becomes a low-pressure zone, while the air outside the airstream (labelled H) stays at higher pressure. The higher-pressure air pushes inwards toward the balloon, forcing it to stay inside the airstream.

At the same time, two other forces are acting:

- Gravity pulls the balloon down.
- The fast-moving air from the dryer pushes it up.

The balloon "floats" because these forces balance — the upward push of the low-pressure airstream against gravity, and the sideways push of higher-pressure air keeping it in place.

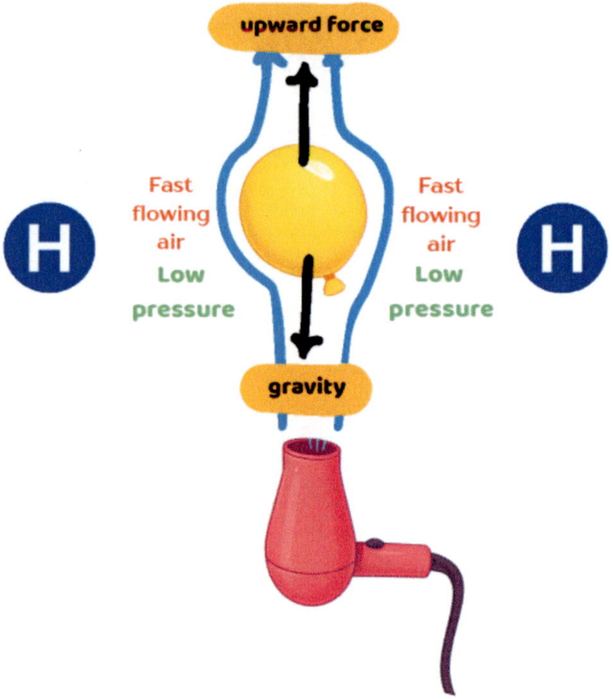

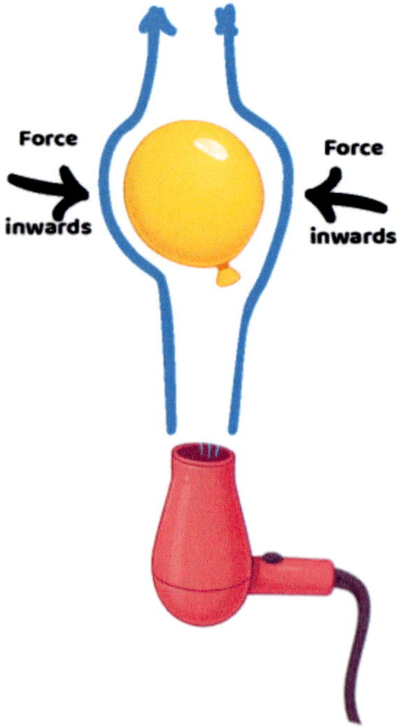

How Far Can You Move the Balloon?

Angled Air Challenge!

Once the balloon is floating, slowly tilt the hair dryer at different angles. Can you keep the balloon balanced in the airstream while moving it to the side? See how far you can guide it without letting it fall!

S — Test balloons of different sizes or materials (latex, foil, rubber). Do they float differently in the airstream? Record which type is most stable.

T — Use a phone or tablet to record the balloon in slow motion. Watch how it wobbles when you tilt the hair dryer — can you see the invisible air pushing it back into place?

E — Build a "balloon stabiliser" by designing cardboard wings or guides that help the balloon stay in the airstream longer. Test and redesign your invention.

A — Create a cartoon or comic of the balloon's "air adventure." Give the balloon a face and show its journey as it battles gravity and zooms through the airstream!

M — Measure how long the balloon stays floating at different hair dryer angles. Create a table to display your data.

Portal Malfunction: The Fart Cloud Nebula

"ALERT! TEMPORAL-BIOGAS CROSSWIRE DETECTED!"
"ENVIRONMENT CLASSIFIED: TOXIC. CLOUDY. EXTREMELY WINDY."

It begins with a sharp ZZZZT! — like a lightning bolt trapped inside a tin can.

The lights in the Secret Lab flicker. The portal stutters. Reality... bends and warps.

A sideways blast of wind WHIPS through the room, sucking up papers, pencils, and Picklebottom's lunch wrapper.

Dr Puddledrip yelps, grabbing a desk leg.
"Hold on! The portal's reversing the airflow!"

A deep FWOMP! rattles the floor. Colours smear. The lab tilts. Everything lurches sharply sideways—

And then— WHOOMPH-CRACKLE-PLOOP!
The world tears open like a curtain caught in a cyclone... and the Secret Lab vanishes.

Everything becomes green.

Swirling.
Weightless.
Cloudy.
Suspiciously... bubbly.

The floor disappears. The ceiling disappears. Everything becomes gas.
Dr Puddledrip steadies herself mid-air.
"Oh dear. This doesn't look like any biological chamber I recognise."

Professor Picklebottom floats past, tumbling in slow motion like a confused balloon.
"Excellent news! We've entered a fart dimension!"

A massive methane bubble drifts lazily beside them — then POP! A shockwave rattles the air.

Puddledrip groans. "Oh no... this isn't just any fart."

She taps frantically on her holo-tablet. "Junior scientists… brace yourselves. We've landed in a crossed temporal stream. These gases aren't from today — they're prehistoric."

Picklebottom's eyes widen. "You mean… time-travel gas?" Puddledrip nods.
"Yes. This cloud is the by-product of a woolly mammoth's digestive emergency."

A deep rumble shakes the nebula. The gas swirls violently around them, fizzing and burbling like a giant, angry soda. Picklebottom sniffs the air.
"…Ah. I recognise this aroma. Grass… wildflowers… and just a hint of sulfur."

He grins." Definitely mammoth."

**Complete the next experiment to disperse the gas cloud before the portal SLAMS SHUT!"
The fate of the Secret Lab depends on it… and on your nose.**

Newton's Wacky Apple Challenge

Discover how gravity, friction, and inertia pull the strings in this topsy-turvy challenge!

They say one falling apple helped Newton unlock the secrets of gravity. But what ELSE was going on? With friction, inertia, and gravity all competing for control, it's time for YOU to take charge and see if you can predict where the apple will land!

Portal Progress

Before we start — what do YOU think will happen?

- **Will the apple shoot sideways when you pull the card?**
- **Or will it drop straight down into the cup?**

What You Need

- 1 large water pitcher (filled ¾ with water)
- 1 sturdy rectangular card
- 1 long cardboard roll (like a paper towel tube — to balance the apple on)
- Test objects: apple, orange, lemon, or even an egg!
- A table or flat surface for setup
- Parent or adult supervision (especially for splashes!)

Step-by-Step:

1. Balance the Cardboard Tube

- Place the card on top of the jug of water. Stand the cardboard tube straight up in the centre of the card.

2. Get Ready to Drop

- Hold the apple (or orange) above the tube so it's lined up with the jug's opening.

3. Let Go!

- Quickly pull the card sideways. The card zips out – and the apple drops straight down into the jug!

Tip: Tip: If it misses, try again. The faster and straighter you pull the card, the better it works.

Observations

- **Did the apple drop straight down into the water?**
- **If you film it in slow motion, you might spot the apple pausing in mid-air for a split second before it falls.**

Safety

Adult supervision required.
Water can splash onto the table or floor. Wipe up spills right away so no one slips.

What's Going On?

According to Newton's First Law of Motion, an object at rest will stay at rest unless acted upon by an unbalanced force.

The apple was resting on top of the cardboard tube. When the card was pulled sideways, the friction between the tube and the card made the tube zip out quickly from underneath the apple.

Because of inertia, the apple resisted the change in motion and stayed in the same place for a split second. But once the tube was gone, there was no longer any upward force holding it up. The forces became unbalanced – gravity was the only force acting – so the apple dropped straight down into the water.

Does the mass of the object affect how it falls?

Try the experiment again using different objects – a lemon, orange, egg, or golf ball. Watch carefully: do they all fall the same way?

In science, it's important to change only one thing at a time (this is called the independent variable). In this case, the thing you change is the object's mass. Everything else – the card, tube, and jug of water – stays the same so you can clearly see the effect of mass on how the object falls.

Take it Further with Exclusive Parent & Teacher Resources

We explored gravity, inertia, friction, and Newton's First Law of Motion in this experiment — ideas that can be tricky for kids to fully grasp.

Next-Level Science: Inside our Secret Lab, we show you variations of this experiment — plus extra ways to teach and extend it.

You'll get: **www.thecrazyscientistlab.com.au**

- Explanations of this experiment & Newton's 1st Law of Motion.
- Hypothesis, variables & more.
- Teacher notes comparing physical vs chemical changes.

S	Test different objects (apple, orange, lemon, golf ball, egg). Do heavier or lighter objects fall differently when the tube is pulled away? Record your results.
T	Use a phone or tablet to film the moment in slow motion. Can you see the exact instant the tube slides and the object begins to fall?
E	Design a "tube launcher" or slider that lets the card/tube be pulled away quickly and smoothly every time. Does it make your results more reliable?
A	Draw or create a cartoon of the apple's "drop journey" — show the forces at work (gravity pulling down, support force gone). Give it a funny thought bubble!
M	Measure and compare the time it takes different objects to fall. Make a chart or graph showing whether mass affects falling speed.

Mission 18

The Force of the Sauce

Let's explore how pressure, buoyancy, and a squeeze of Jedi-level power let YOU control objects inside a bottle — just like a real mind trick!

Portal Progress

It may look like a ketchup packet lazily floating on the surface... but with a simple squeeze of the bottle you can command it like a Jedi mind trick! When you squeeze, the pressure increases, the packet's density changes — and suddenly you've got the power to make your ketchup submarine sink, rise, or hover on command!

Before we start — what do YOU think will happen?"

- **Will the ketchup dive straight to the bottom when you squeeze?**
- **Or will it hover in the middle — calmly waiting for your next Jedi command?**

What You Need

- 1 sauce packet (ketchup, soy, mustard — any small sealed packet works)
- 1.25 L clear plastic bottle filled with water + its cap
- Bowl of water (for testing packet buoyancy)

Take it Further

Try attaching a tiny weight (pen lid or Blue-Tac) to the packet to adjust how easily it sinks. Can you fine-tune the perfect "hover mode"?

Step-by-Step:

1. Choose Your Padawan Packet

Drop the sauce/ketchup/mustard packets into a bowl of water.

- Floats? Knighted as a Diver.
- Sinks? Send it back to the Dark Side and choose a new recruit.

2. Fill Your Training Bottle

- Fill a clear bottle right to the top with water (no gaps).
- Seal it super-tight — even a tiny air pocket makes the trick harder.

Professor tip: Less air in the bottle = better "squeezability" and smoother diving.

3. Insert the Diver

- Slide your floating packet into the bottle.
- Cap it tightly (no air gap!)

Your Diver should bob near the surface like a tiny tomato raft.

4. Use the Force (aka The Squeeze)

- Squeeze = packet dives to the bottom.
- Relax = it floats back up.
- Gentle squeeze = hover mid-water.

Advanced Jedi move: hold it perfectly still in the middle for 3 seconds.

Fine-Tune Like a Scientist

- It won't dive: Add a tiny pinch of salt, recap, and try again. (Salt makes the water denser, so the packet needs less squeezing to sink.)
- It sinks all the time: Pour out a little water, add fresh plain water back to the very top, recap tight, and test again. Try a different packet if needed.
- Bubbles on the packet? Tap the bottle to knock them off—bubbles can mess with buoyancy.

What's Going On?

Inside your sauce packet is a tiny bubble of air. When the bottle is just sitting there, that bubble is big, so the packet is less dense than the water – it floats.

But when you squeeze the bottle, you increase the pressure on the packet. The extra pressure squashes the air bubble, making it smaller. Now the packet becomes denser than the water – and suddenly it sinks!

Release the squeeze, and the bubble expands again. The density goes back down – whoosh! The packet floats up again.

Jedi Trick: By controlling the size of the air bubble with your "Force squeeze," you decide whether the packet rises, falls, or hovers in the middle!

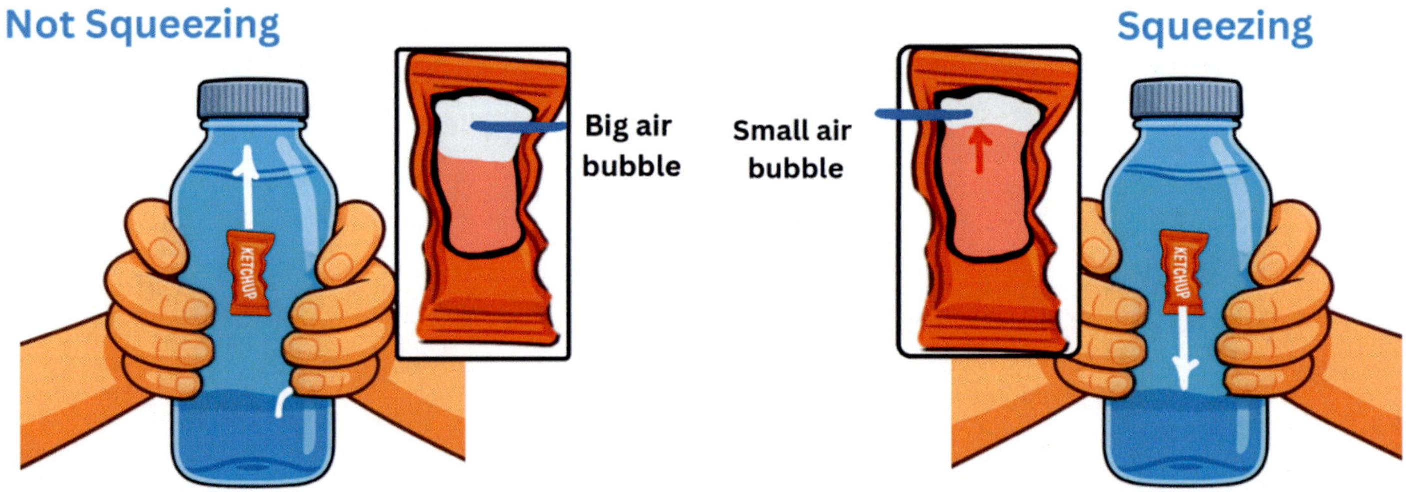

Take It Further: The Clear Pen Lid Diver

Want to see the "Force" in action even more clearly? Try swapping your sauce packet for a clear plastic pen lid.

Seal the open end with a little Blue Tack or plasticine so it traps some air.
Drop it into your training bottle. If it floats, you're ready to go Jedi-mode!
Now when you squeeze the bottle, you can actually see the air bubble shrink and expand inside the lid.
This makes it super easy to understand why the diver sinks (more dense) and floats (less dense).
You can even compare different amounts of air in the lid to see how it changes the trick!

Fishy Fact!

Fish have a special organ called a swim bladder - it works just like your ketchup packet! By squeezing or relaxing tiny muscles, and also adding oxygen into or out they change the size of the air inside.
Bigger bubble = float.
Smaller bubble = sink.
That's how fish cruise up and down in the water without swimming

S Test different divers — ketchup, soy sauce, mustard, or even the clear pen-lid diver. Which ones sink faster? Which ones hover most easily in the middle?

T Film your Jedi squeeze in slow motion on a phone or tablet. Can you capture the exact moment the air bubble shrinks and the diver sinks?

E Build your own "diver lab"! Try bottles of different shapes or materials. Does a wider bottle make it harder or easier to squeeze — and easier or harder to control the diver?

A Draw a cartoon of your ketchup diver as a tiny submarine with a pilot inside. Give them a thought bubble when you "Force-squeeze" them down into the depths!

M Count how many squeezes it takes to keep your diver hovering for 10 seconds. Compare with a friend — whose Jedi powers are stronger? Then record and graph your results.

Eggbert the Expanding Chicken!

Ever wondered what happens when air gets warm, cold, or fizzy — all at the same time?
Today we're bringing Eggbert the Expanding Chicken to life using three bottles, a few bubbles, and a whole lot of clucking science!
You'll see how gases form in a reaction, expand when heated, and shrink when cooled — all without a single fried feather.
Grab your goggles and get ready for some bubbling poultry pressure!

Portal Progress

Which bottle will make Eggbert grow the biggest head?

- **What will happen when we heat the gas?**
- **What will happen when we cool it down?**

Write down your guesses before Eggbert starts inflating!

What You Need

- 3 clear plastic bottles (same size bottles)
- 3 latex gloves or balloons (Eggbert heads!)
- 3 Alka-Seltzer tablets (or 1 tsp bicarb + 2 tbsp vinegar)
- Warm water (in a jug)
- Cold water (in a bowl or fridge)
- Room-temperature water
- 2 tubs/bowls for hot and cold water baths
- Permanent marker (label "HOT," "COLD," "ROOM")
- Tray, goggles, paper towels
- Sticky tape

Step-by-Step:

1. Get your flock ready

- Pour ¼ cup of warm water into each clear plastic bottle.
- Drop 1 Alka-Seltzer tablet into each.
- Quickly stretch a glove or balloon over the top —
- Eggbert's head is on! Tape the glove so no gas can escape.

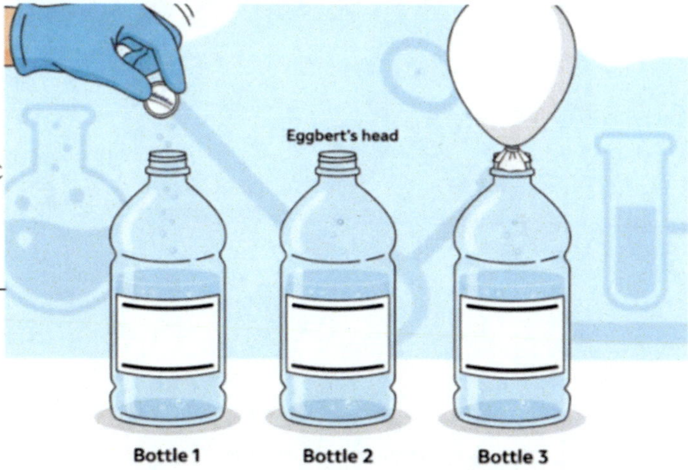

2. Split your flock

- Place one bottle in a bowl of hot water.
- Place another in a bowl of cold water (or in the fridge).
- Leave the third at room temperature.

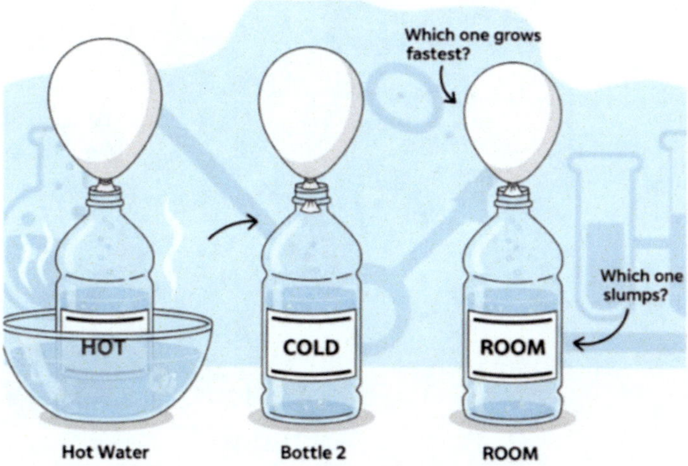

3. Observe and record

- Use a ruler to measure each glove's height, or draw what you see.

Note the order of Eggbert heads from biggest → smallest!

Dr Puddledrip's Breakdown

Gases are made of tiny, zooming particles that spread out to fill any space they can find.
That's exactly what happens inside Eggbert the Expanding Chicken — the gas takes up space, pushes on the glove, and puffs him up!
But the story doesn't stop there...

Cold | Hot

Particles closer together | **Particles spread out**

When Eggbert gets hot, those gas particles absorb energy and move even faster — they bump around, spread out, and make his head grow even bigger.
When he gets cold, the particles slow down as they lose energy and snuggle closer together — so his head shrinks back down!
That's science in action: adding energy means expansion; removing energy means contraction.
No feathers required!

Inside the Bottles

1 Chemical Reaction

Alka-Seltzer + water → fizzing bubbles of carbon dioxide gas (CO_2).

These bubbles fill the space above the water and stretch the glove.

2 Hot Water Eggbert

The warm bath adds energy to the gas.

Particles move faster, push harder, and make Eggbert's glove expand more.

3 Cold Water Eggbert

Cooling removes energy.

Particles slow down and move closer, so the gas takes up less space and the glove deflates.

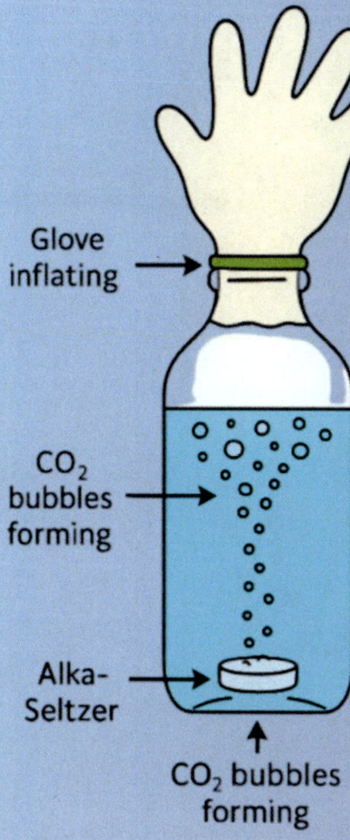

S You've seen what happens when Eggbert gets hot or cold — now test what happens when you change something else!
Try using different amounts of Alka-Seltzer tablets or changing the depth of the water.

T Record Eggbert's reactions using a tablet or phone. Create a time-lapse video that shows how the glove inflates and deflates — can you spot the moment the reaction slows down?

E Design a "Glove Stabiliser" stand using cardboard, LEGO, or a recycled bottle holder. Make sure Eggbert stands upright while he expands!

A Draw Eggbert's journey as a comic!
Show the bubbles forming inside, the glove puffing up, and Dr Puddledrip cheering him on.

M Measure the glove's height before and after heating and cooling. Then calculate how much it changed — can you work out the percentage change in volume?

Mission 20

Milo the Magical Mason Jar

Ever wondered what happens when a liquid gets toasty warm and its molecules start stretching like they're doing morning yoga?
Today we're turning an ordinary jar into a science superstar — Milo the Magical Mason Jar!

Fill him to the brim, stick in a straw, and watch in amazement as the coloured water climbs up the straw all by itself when the jar gets hot. No spells, no bubbles — just pure expanding science magic!
Grab your goggles and get ready to witness what happens when liquids get a little steamy!

Portal Progress

Before You Begin — Make a Prediction!

- **What will happen when we heat it?**
- **• What will happen when it cools down again?**
- **• Write down your guesses before Milo starts stretching!**

What You Need

- 1 glass mason jar (or any clear jar with a lid)
- 1 drinking straw
- Food colouring
- Plasticine, Blu Tack, or modelling clay
- Warm water and cold water
- A bowl or jug for very hot or boiling water (ask an adult helper!)
- Tray or towel (to catch spills)

Step-by-Step:

1. Fill and Seal:

- Fill your jar right to the top with coloured water.
- Push the straw through the lid and seal around it with Plasticine so no air can escape.
- The straw should poke a 10 centimetres above the lid.

Now you've made your very own Liquid Expansion Detector 3000!

2. Heat it Up!

- Carefully place the jar into a bowl of hot water (ask an adult to help with boiling water).
- Watch closely — does the water level change?

Use. a marker to make a mark at the highest point.

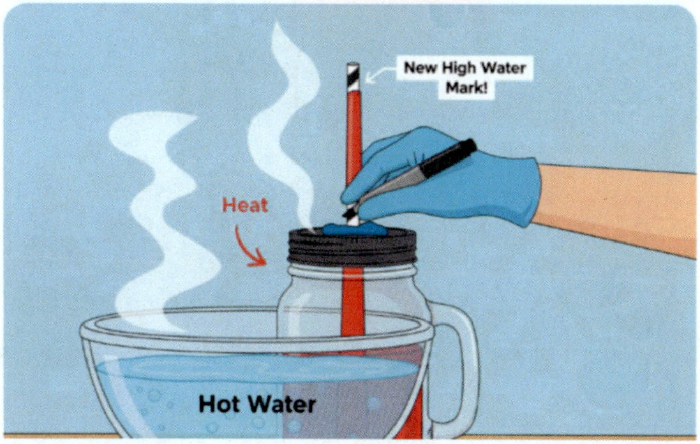

3. Cool it Down!

- Move the jar into a bowl of cold water, or place it in the fridge for a few minutes.
- What happens to the water level now?

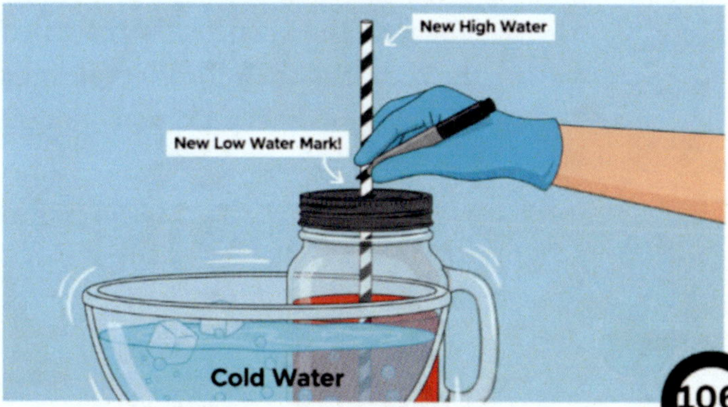

Professor Pickelbottom's Breakdown

Liquids might look calm, but inside they're full of tiny, jiggling particles that never stop moving! When those particles warm up, they get a burst of energy, start zooming around faster, and push each other further apart — that's why the liquid rises up the straw!

Cool things down again, and those same particles lose energy, snuggle closer together, and the liquid shrinks back down.

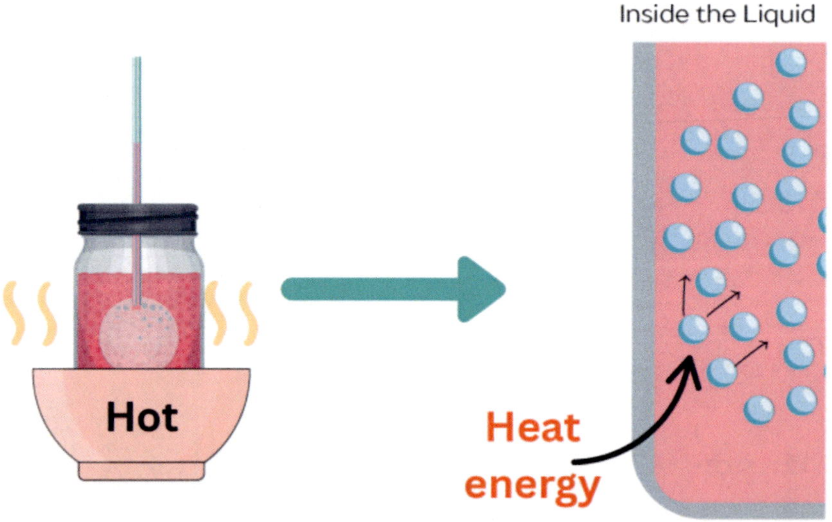

Heating Milo

When Milo is placed in warm water, heat energy flows into the jar. This makes the particles in the coloured water move faster and spread out. As the particles spread out, they take up more space — that's why the liquid expands and rises up the straw!

Cooling Milo results in energy leaving the water in the jar and moving into the cold water around it. This means the particles vibrate less and take up less space. The water level goes down in the straw.

How a Thermometer Works

When it gets hotter, the liquid inside the thermometer expands and rises up the tube. When it gets colder, the liquid contracts and falls back down.

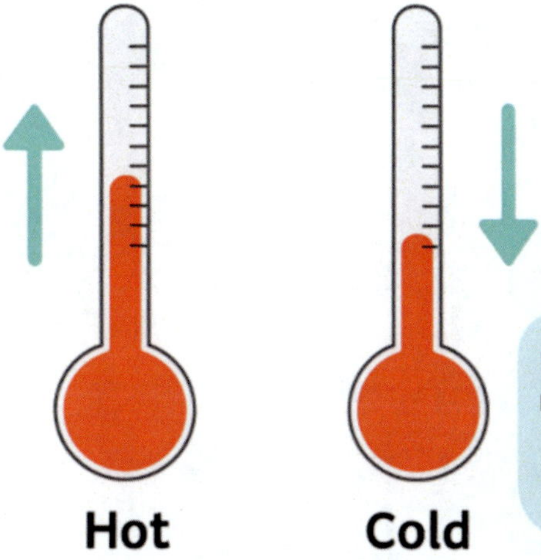

Did You Know?
Liquid thermometers contain alcohol, which is often dyed red or blue.

S Try different liquids! Does oil expand more or less than water? What about juice or milk?

T Film your experiment close-up to capture the water level moving in the straw. Create a time-lapse or slow-motion clip showing the water rising and falling as it heats and cools.

E Use extra straws to build a mini "Science Stand" that supports your jar like a tower! Test different shapes — triangles, squares, or pyramids — to see which structure stays strongest when the hot and cold water make things shift and wobble.

A Add speech bubbles or comic labels like "I'm heating up!" or "Cooling down!"

M Calculate the difference between the two levels — can you work out the percentage change?

Back to the Lab?!

There is a BOOMPF!
A WHOOSH!
And then—THUD!

Dr Puddledrip and Professor Picklebottom drop out of a swirling tear in space-time and land in a tangled heap on the Secret Lab floor. Papers flutter down like confetti. A beaker rolls across the tiles. Someone's leftover sandwich lands on Picklebottom's head.

He sighs.
"Well... at least it isn't mammoth-flavoured this time."

Puddledrip scrambles upright, brushing snow, fumes, and suspicious green sparkles off her lab coat.
"We made it! We're back in the—"

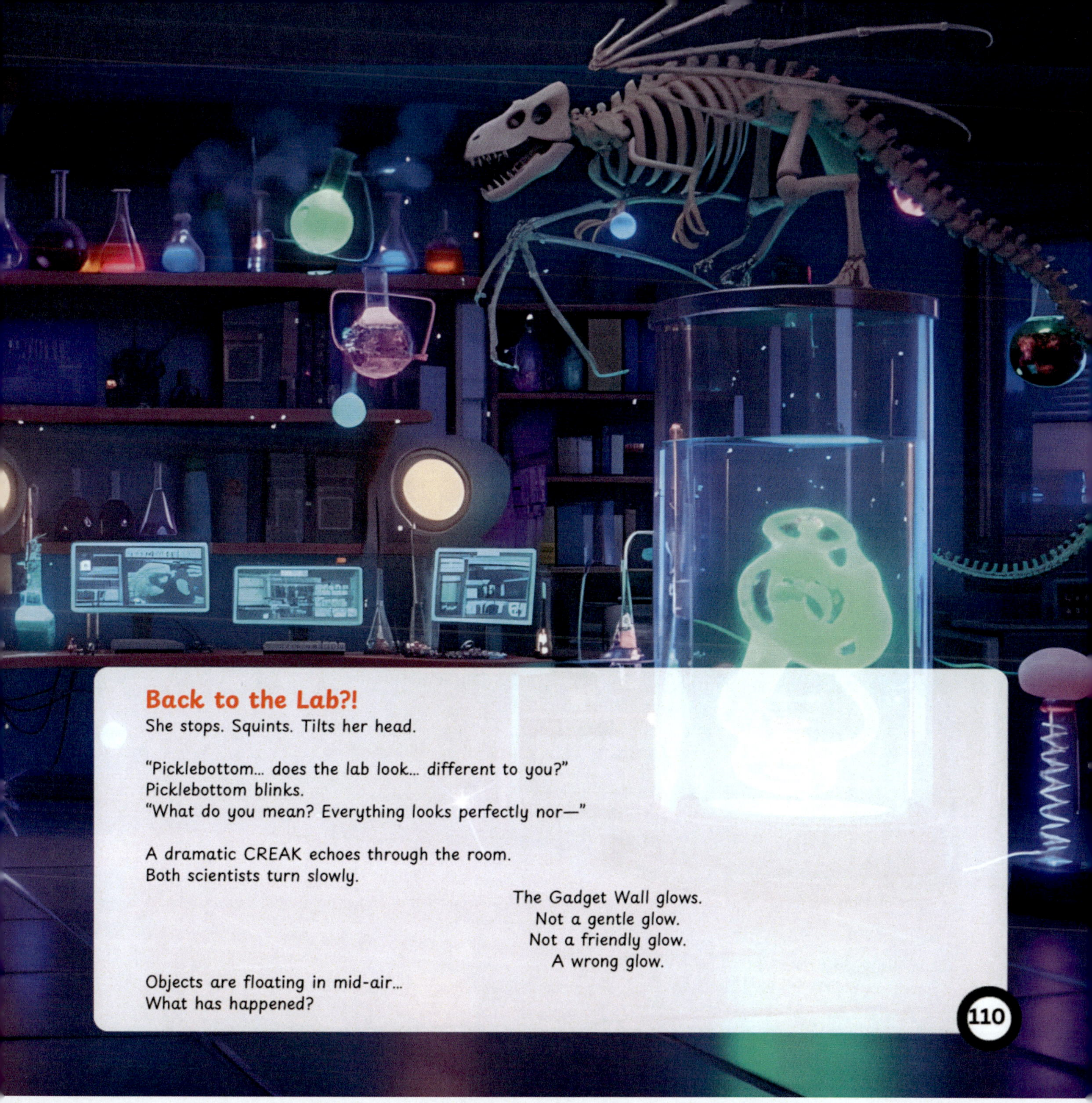

Back to the Lab?!

She stops. Squints. Tilts her head.

"Picklebottom… does the lab look… different to you?"
Picklebottom blinks.
"What do you mean? Everything looks perfectly nor—"

A dramatic CREAK echoes through the room.
Both scientists turn slowly.

The Gadget Wall glows.
Not a gentle glow.
Not a friendly glow.
A wrong glow.

Objects are floating in mid-air…
What has happened?

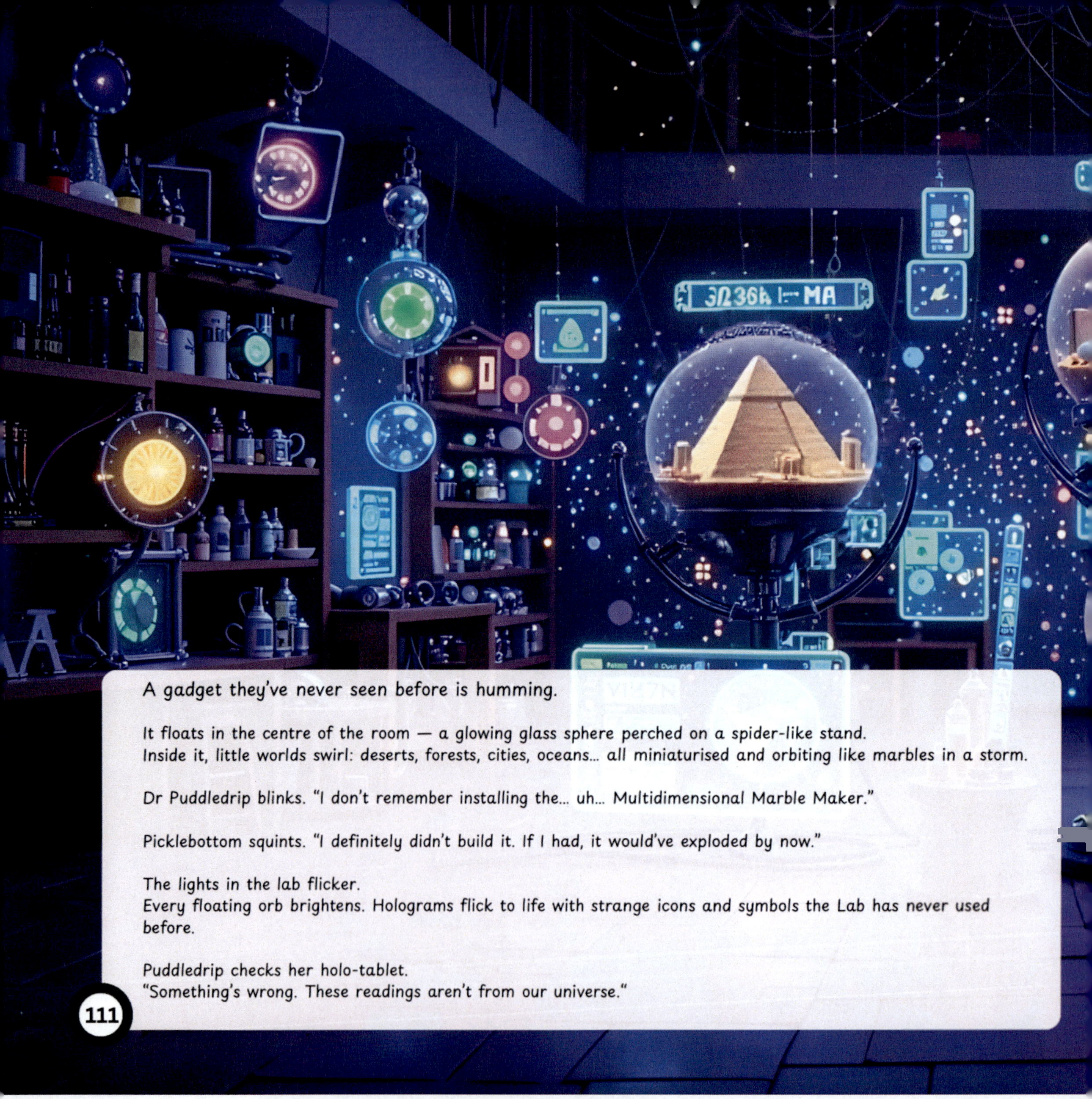

A gadget they've never seen before is humming.

It floats in the centre of the room — a glowing glass sphere perched on a spider-like stand.
Inside it, little worlds swirl: deserts, forests, cities, oceans… all miniaturised and orbiting like marbles in a storm.

Dr Puddledrip blinks. "I don't remember installing the… uh… Multidimensional Marble Maker."

Picklebottom squints. "I definitely didn't build it. If I had, it would've exploded by now."

The lights in the lab flicker.
Every floating orb brightens. Holograms flick to life with strange icons and symbols the Lab has never used before.

Puddledrip checks her holo-tablet.
"Something's wrong. These readings aren't from our universe."

The centre sphere makes a soft pop.

Then a louder POP POP POP. Then—

WHOOOOOOOMP!

A beam of shimmering light fires from the main globe, wrapping around Picklebottom like a lasso.

“PUDDLEDRIP! I’M BEING BUBBLED!” he yelps as he’s yanked off the floor.

The beam snaps toward Puddledrip next.
“Oh come on— I haven’t even had lunch!” she cries as she’s sucked upward.

Both scientists swirl toward the glowing orb like spaghetti being slurped into another dimension.

"Junior scientists—!!"
"Tell our plant collection we loved them—!!"

ZIIIIP!

The sphere seals shut.

All the globes dim.

The lab falls silent... except for one tiny hologram blinking on the front of the machine.

It reads:

SYSTEM UPDATE COMPLETE.
NEXT DESTINATION: EARTH... UNZIPPED.

What new world have they been pulled into?
Who built the mysterious globes?
And why is Earth coming... undone?

Find out in..

SECRET LAB SCIENCE SERIES: BOOK 2 — EARTH UNZIPPED

coming soon!

THE SECRET LAB FIELD DICTIONARY

- Chem
- General Science
- Physics
- Biology
- Engineering

Absorb

IN THIS BOOK:

The coloured water in Milo the Magical Mason Jar absorbs heat energy from the hot water bath and starts to rise up the straw.

What it means?

When something takes in another substance or energy, like a sponge soaking up water or a surface taking in light or heat.

Adehsion

IN THIS BOOK:

Bubble juice clings to the table and straw in Super Soap Domes, helping the bubble walls spread and hold their shape.

What it means?

The way different substances stick to each other, like water drops clinging to glass or tape sticking to paper.

Water drop + Glass   Tape + Paper

THE SECRET LAB FIELD DICTIONARY

- Chem
- Physics
- Engineering
- General Science
- Biology

Air pressure

IN THIS BOOK:

You use changing air pressure to make a Cyclone in a Bottle and to keep the balloon hovering in Minion Levitation.

What it means?

The push of air on everything around us, caused by air particles bumping into surfaces. Even though we can't see it, this pressure is always there.

BALANCE POINT

The spot where forces are equal so an object can stay steady instead of tipping, falling, or shooting away.

IN THIS BOOK:

In the minion experiment, gravity pulls the ball down while the air from the hair dryer pushes it up.

THE SECRET LAB FIELD DICTIONARY

- Chem
- Physics
- Engineering
- General Science
- Biology

Boiling

IN THIS BOOK:

When popcorn kernels are heated in Steam-Powered Snack Attack, the water inside them turns into steam gas. This isn't boiling in a pot — but it's the same idea: heat gives water energy until it changes into a gas.

What it means?

When a liquid gets hot enough to turn into gas (like water turning into steam).

Buoyancy

IN THIS BOOK:

You explore buoyancy in The Force of the Sauce when squeezing the bottle makes the ketchup packet sink, float, or hover.

What it means?

The upward force that pushes objects up in water or air, helping things float.

THE SECRET LAB FIELD DICTIONARY

Chem
Physics
Engineering
General Science
Biology

Carbon Dioxide

In this book:

Carbon dioxide gas makes Eggbert the Expanding Chicken swell up when the fizzy tablet reacts with water. You also see CO_2 bubbles lifting blobs in the Lava Lamp Lab.

What it means?

A colourless gas we can't see, made when things burn or when fizzy tablets react with water. CO_2 gas is lighter than water, so the bubbles rise to the top — then fall again when the bubbles pop

Chemical Change

"When substances mix and make something new that wasn't there before — like fizzing bubbles, new smells, new colours, or unexpected heat.

In this book:

When you drop an Alka-Seltzer into water in Eggbert or the Lava Lamp Lab, a brand-new gas forms. That's a chemical change — something new is created.

THE SECRET LAB FIELD DICTIONARY

- Chem
- General Science
- Physics
- Biology
- Engineering

What it means?

Chemical Reaction

IN THIS BOOK:

Eggbert the Expanding Chicken inflates because Alka-Seltzer + water react to form carbon dioxide gas.

When substances mix and change into something new — like tablets fizzing or baking soda and vinegar bubbling.

Condensation

IN THIS BOOK:

You see condensation when warm water vapour inside the Mason Jar Heat Rise experiment cools down on the glass.

When warm water vapour (invisible gas) cools down, it changes back into liquid water. This is why droplets appear on cold glass, bathroom mirrors after a shower, or clouds in the sky.

THE SECRET LAB FIELD DICTIONARY

- Chem
- General Science
- Physics
- Biology
- Engineering

Conduction

IN THIS BOOK:

You explore conduction when heat from warm water travels through your container during experiments like the Hot Water Bath or Mason Jar Heat Rise.

What it means?

The way heat travels through a solid material — like a metal spoon getting hot when one end sits in warm water.

CONDUCTOR

A material that lets heat travel through it easily — like metal, copper, or even water.

IN THIS BOOK:

Metal tools in your experiments heat up quickly because they are good conductors of heat.

THE SECRET LAB FIELD DICTIONARY

- Chem
- General Science
- Physics
- Biology
- Engineering

Convection

IN THIS BOOK:

You explore convection in later missions.

What it means?

The movement of warm fluids (liquids or gases) rising and cooler fluids sinking, creating a circulating flow called a convection current.

Density

IN THIS BOOK:

Density explains why the blobs in your Lava Lamp Lab rise and fall. The water, oil, and fizzy-tablet bubbles all have different densities, so they layer instead of mixing.

What it means?

How tightly packed the particles are inside a substance. Objects or liquids that are heavier for their size have a higher density and sink. Things that are lighter for their size have a lower density and float.

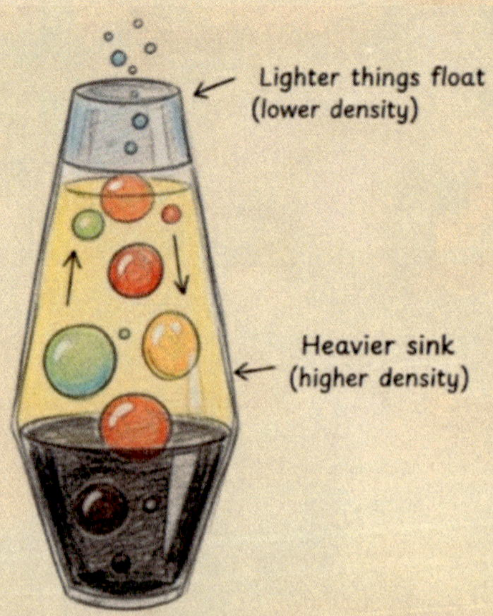

THE SECRET LAB FIELD DICTIONARY

- Chem
- Physics
- Engineering
- General Science
- Biology

Dissolve

IN THIS BOOK:

When you make the Lava Lamp Lab, the fizzy tablet dissolves in the water. The tiny tablet pieces disappear into the liquid and release gas bubbles that rise through the oil.

What it means?

When a solid breaks apart so completely in a liquid that it seems to vanish — it mixes evenly through the liquid.

DRAG

A force that pushes against something moving through a fluid (air or water), making it harder to move — like pushing your hand through water or how a parachute slows a fall

IN THIS BOOK:

When objects fall in Newton's Apple or move through the air in other missions, the air pushes back against them. This push-back is air resistance, a type of drag that slows things down.

THE SECRET LAB FIELD DICTIONARY

- Chem
- Physics
- Engineering
- General Science
- Biology

Elastic

IN THIS BOOK:

You see elasticity when the balloon in Eggbert the Expanding Chicken stretches as carbon-dioxide gas fills it. The stretchy latex lets the balloon expand without bursting — and it snaps back to its original shape when it deflates. That's elasticity at work.

What it means?

How much something can stretch and then return to its original shape — like rubber bands, balloons, or stretchy slime.

Energy

IN THIS BOOK:

Energy makes everything happen — including the moving air that lifts the balloon in Minion Levitation.

What it means?

The ability to make things move, heat up, light up, or change.

Energy comes in many forms — like heat, movement, and sound.

THE SECRET LAB FIELD DICTIONARY

Chem
General Science
Physics
Biology
Engineering

Evaporation

IN THIS BOOK:

Seen in a future misson.

What it means?

When a liquid slowly turns into a gas — like puddles shrinking in the sun or warm water slowly disappearing into the air.

EXPANSION

When something spreads out or takes up more space — like heated air getting bigger, gas filling a balloon, or bubbles growing as fizz forms.

IN THIS BOOK:

In Fizztastic Flying Fiasco, gas builds up inside the film canister. As more gas forms, it expands — pushing harder and harder until the lid blasts off!

THE SECRET LAB FIELD DICTIONARY

- Chem
- General Science
- Physics
- Biology
- Engineering

Experiment

IN THIS BOOK:

Every mission in this book is an experiment. You're testing an idea to discover a result!

What it means?

A planned test you do to find out what happens. You follow steps, watch closely, and record what occurs.

Fair Test

IN THIS BOOK:

In The Big Fizz-Off, you compare different fizzy fuels. To make it a fair test, you use the same amount of baking soda, the same bottle size, and the same time — the only thing that changes is the liquid inside.

A test where you change only one thing and keep everything else the same, so you can compare the results fairly.

THE SECRET LAB FIELD DICTIONARY

- Chem
- General Science
- Physics
- Biology
- Engineering

Force

IN THIS BOOK:

When your Fizz-Tastic Flying Fiasco rocket pops its lid and shoots into the air,the gas pushes on the lid and the rocket — sending the lid down and the rocket up.

What it means?

A force is a push or a pull that can make something start moving, stop, speed up, slow down, or change direction.

FRICTION

Friction is a force that happens when two surfaces rub together. It usually slows things down or stops them.

IN THIS BOOK:

When an object drops in Newton's Wacky Apple Challenge or slides along a surface, friction with the air and the ground helps slow it down so it doesn't keep moving forever.

THE SECRET LAB FIELD DICTIONARY

Chem
General Science
Physics
Biology
Engineering

Gas

What it means?

A gas is a state of matter where particles are far apart and moving freely. Gases spread out to fill any space they're in are often invisible, like air.

IN THIS BOOK:

In Kaboom-in-a-Bag and the Mini Firefighter, reactions make carbon dioxide gas (CO_2). You can't see the gas itself, but you do see it inflate bags and balloons or push out flames.

Gravity

Gravity is the invisible pull that draws objects toward each other. On Earth, gravity pulls everything toward the centre of Earth — which we feel as 'down'.

IN THIS BOOK:

In Newton's Wacky Apple Challenge, gravity is what makes the apple (and everything else) fall down instead of up. Your rockets and jumping raisins all come back down because Earth's gravity is always pulling on them.

THE SECRET LAB FIELD DICTIONARY

Chem
General Science
Physics
Biology
Engineering

Heat

In this book:

In The Great Popcorn Explosion, heat made the water inside each popcorn kernel turn to steam… until POP! the kernel burst open.

What it means?

Heat is a type of energy that makes things warmer. When something gains heat, its particles move faster. When it loses heat, its particles slow down.

Inference

Using clues or evidence to make a smart guess about something you can't see directly — like solving a tiny science mystery.

In this book:

You use inference whenever you explain why something happened — like guessing why raisins suddenly start dancing

THE SECRET LAB FIELD DICTIONARY

- Chem
- General Science
- Physics
- Biology
- Engineering

Ingredient

What it means?

A single part that goes into a mixture or reaction — like a recipe ingredient, but for science!

IN THIS BOOK:

Every kitchen-chemistry mission uses ingredients — lava lamp oil, fizzy tablets, corn kernels, salt, bubbles, vinegar, baking soda, detergent, food colouring and more.

LIQUID

A state of matter that flows, takes the shape of its container, and isn't solid or gas — like water, oil, or vinegar.

IN THIS BOOK:

Liquids appear everywhere: Lava Lamp Lunacy (oil + water), Density Tower (liquid layering) and more.

THE SECRET LAB FIELD DICTIONARY

- Chem
- General Science
- Physics
- Biology
- Engineering

Mantle

IN THIS BOOK:

We don't explore Earth's mantle directly in Book 1but it appears in the story world when you use the Lava Lamp Lab to imagine how hot blobs rise and cooler blobs fall — just like convection inside Earth's mantle.

What it means?

The thick, hot, rocky layer beneath Earth's crust. Slow, swirling movement in the mantle pushes Earth's plates around — causing earthquakes, volcanoes, and continents to drift.

MATTER

Anything that takes up space and has weight. Matter can be a solid, liquid, or gas.

IN THIS BOOK:

Every experiment uses matter — solids (corn kernels, tablets), liquids (water, oil), and gases (steam, carbon dioxide). Matter is everywhere in each mission.

THE SECRET LAB FIELD DICTIONARY

Chem
General Science
Physics
Biology
Engineering

Melting

IN THIS BOOK:
You see melting in Marshmallow Monsters when heat makes the gooey sugar-gelatin walls soften and lose their shape.

What it means?

When a solid turns into a liquid because it gets warmer — like ice melting into water or chocolate softening in your hand.

Molecule

Tiny particles too small to see. Almost everything around you — water, air, and even you — is made of molecules.

IN THIS BOOK:

Fizz tablets, vinegar, baking soda, and even water are made of molecules that change or rearrange during your experiments.

THE SECRET LAB FIELD DICTIONARY

Chem
General Science
Physics
Biology
Engineering

Observation

IN THIS BOOK:

You make observations in every mission — watching bubbles rise, kernels explode, colours swirl, balloons expand, and rockets launch.

What it means?

Using your senses (seeing, hearing, touching, sometimes smelling) to notice what is happening during an experiment

OUTER CORE

The hot, liquid metal layer deep inside the Earth, found beneath the mantle. It's made mostly of melted iron and nickel.

THE SECRET LAB FIELD DICTIONARY

Chem
General Science
Physics
Biology
Engineering

Oxygen

What it means?

A gas in the air that living things need to breathe. Oxygen helps fires burn, balloons inflate, and chemical reactions happen. We can't see it, but it's all around us.

IN THIS BOOK:

You meet oxygen when bubbles form in fizzing experiments (like Eggbert or the Lava Lamp Lab). Oxygen is also part of the air that fills balloons, helps things burn, and keeps us alive.

Particle

The teeny-tiny pieces that everything is made of — too small to see without a microscope.

IN THIS BOOK:

Particles show up in Steam-Powered Snack Attack, Marshmallow Monsters, and Lava Lamp Lunacy — where tiny bits of matter move faster when heated or slower when cooled.

THE SECRET LAB FIELD DICTIONARY

- Chem
- General Science
- Physics
- Biology
- Engineering

Physical Change

What it means?

A change in size, shape, or state (solid → liquid → gas) where nothing new is made.

IN THIS BOOK:

When marshmallows shrink or puff (heat + cool), or when ice melts in the Hot Water Bath, those are physical changes.

PREDICTION

A prediction is a smart science guess. Before you start an experiment, think about what you think will happen. Write or draw your prediction — then test it and see if you were right!

IN THIS BOOK:

You'll make a prediction before every crazy mission. Guess how high the Fizz-tastic Flying Fiasco rocket will fly, how huge Eggbert the Expanding Chicken will grow, and whether the Minion Levitation balloon will keep hovering.

Then you'll test your ideas like a real scientist.

THE SECRET LAB FIELD DICTIONARY

- Chem
- General Science
- Physics
- Biology
- Engineering

Pressure

IN THIS BOOK:

In Earth in a Jar, we see the outer core modelled, where huge heat and pressure keep the metal liquid.

What it means?

When something pushes on something else. Air, water, and even your feet on the ground all press and push — that push is called pressure.

Reflection

IN THIS BOOK:

Reflection is shown in Mini Firefighter or Cyclone in a Bottle, where shiny water surfaces reflect (bounce back) light.

When light hits a surface and bounces back — like seeing yourself in a mirror.

- Chem
- General Science
- Physics
- Biology
- Engineering

Refraction

IN THIS BOOK:

Seen when looking through water layers in Density Tower or the bottle in Earth in a Bottle — objects appear bent or stretched.

What it means?

When light bends as it moves from one substance to another — like a straw looking broken in water.

THE SECRET LAB FIELD DICTIONARY

Chem
General Science
Physics
Biology
Engineering

Sink

What it means?

To fall or move downward in a fluid because it is heavier (denser) than the fluid around it.

IN THIS BOOK:
You see sinking in The Force of the Sauce, Lava Lamp Lunacy, and Raisin Rave when denser objects fall through a liquid.

Solid

A state of matter that keeps its shape — like a rock, a coin, or a fizzy tablet.

IN THIS BOOK:

You see solids in many missions — like salt, baking soda, and tablet pieces. You can also see how solids seem bent or stretched when you look at them through layered liquids in Density Tower or Earth in a Bottle.

Chem
General Science
Physics
Biology
Engineering

Solubility

What it means?

How easily something dissolves in a liquid. Sugar dissolves easily in water — that's high solubility. Sand does not dissolve — that's low solubility.

SOLUTION

A special mixture where a solute (often a solid) dissolves completely in a solvent (liquid) — like sugar disappearing into water to make sugar water, so it looks the same all the way through.

IN THIS BOOK:

You see a solution in Lava Lamp Lunacy when the colour tablet dissolves into the water.

THE SECRET LAB FIELD DICTIONARY

- Chem
- General Science
- Physics
- Biology
- Engineering

Vortex

IN THIS BOOK:

You create swirling vortexes in Cyclone in a Bottle, and you see mini vortex swirls in Lava Lamp Lunacy and The Force of the Sauce as blobs and bubbles spin downward through the liquid.

What it means?

When a liquid or gas spins in a circle, forming a swirling, funnel-shaped spin called a vortex — like water spiralling down a drain or a mini-tornado inside a bottle.

Weight

Weight is how strongly gravity pulls on an object. Heavier things feel more pull and are harder to lift.

IN THIS BOOK:

Weight affects how objects behave in Raisin Rave, The Force of the Sauce, and Kaboom-in-a-Bag. Heavier objects sink faster, push down harder on surfaces, and fall more quickly.

Want Even MORE Secret Lab Science?

Pssst... Mini-Scientists!

If you enjoyed exploring the science words in this Field Dictionary, you'll love what's waiting inside The Secret Lab Teacher Membership!

Inside the membership you'll find:

- Hundreds of extra science words explained in kid-friendly language
- Printable classroom posters and vocabulary cards
- Step-by-step teacher notes for every experiment
- Bonus mini-lessons and animated explainers
- Extension tasks, literacy links, and curriculum-aligned worksheets
- Perfect for classrooms and curious kids at home!

Unlock even more science superpowers at:

www.thecrazyscientistlab.com.au

Ask your teacher or grown-up to check it out - your learning journey doesn't stop here... The Secret Lab is always open!

TEACHER SECRET LAB

PARENT LAB

Made in United States
Orlando, FL
04 November 2026

90200576R00088